2009 INTERNATIONAL PLUMBING CODE NEED TO KNOW

About the Author

R. Dodge Woodson has written more than 100 non-fiction books. He is a licensed general contractor and a licensed master plumber who has built as many as 60 single-family homes a year. Woodson is a well-known remodeling contractor, plumbing contractor, and business owner. Additionally, Woodson is accredited as an expert witness and serves as a consultant on construction and plumbing litigations. He lives in Brunswick, Maine.

2009 INTERNATIONAL PLUMBING CODE NEED TO KNOW

The 20% of the Code
You Need 80% of the Time

R. Dodge Woodson

New York Chicago San Francisco Lisbon London Madrid
Mexico City Milan New Delhi San Juan Seoul
Singapore Sydney Toronto

The McGraw·Hill Companies

Cataloging-in-Publication Data is on file with the Library of Congress

2009 International Plumbing Code Need to Know

1 2 3 4 5 6 7 8 9 0 FGR/FGR 0 1 4 3 2 1 0 9

ISBN 978- 0-07-154449-8
MHID 0-07-154449-6

Sponsoring Editor
Joy Bramble Oehlkers

Acquisitions Coordinator
Michael Mulcahy

Editorial Supervisor
David E. Fogarty

Project Manager
Jacquie Wallace, Lone Wolf Enterprises, Ltd.

Copy Editor
Jacquie Wallace, Lone Wolf Enterprises, Ltd.

Proofreader
Roger Woodson, Lone Wolf Enterprises, Ltd.

Indexer
Leona Woodson, Lone Wolf Enterprises, Ltd.

Production Supervisor
Pamela A. Pelton

Composition
Lone Wolf Enterprises, Ltd.

Art Director, Cover
Jeff Weeks

Dedication

This book is dedicated to my son, Adam.
I have watched Adam grow and mature into a fine young man
and he continues to impress me with his iron will and strong values.
Adam and Afton are the best children a father could ever hope for.

Contents

Preface

This book is your guide to getting your code questions answered with less stress. It is your ticket to simplifying the cryptic code that you work with on a daily basis. The code is large and can be complicated. I've translated the code into easy-to-use terms for people in the field. See the code as a real-world guide instead of some foreign language that only some people can begin to sift through. This is your chance.

How important is understanding the code? It is essential for professionals in the trade. Building without code compliance is an expensive mistake. Many good builders and contractors mean well and still stray from the code. This is often due to the code being difficult to understand and follow. Not anymore. This book will walk you through the code requirements chapter by chapter.

Laid out in the same basic sequence as the code book, this book can be used in conjunction with your code book. While this book is not a replacement for your code book, it is a strong enhancement to it. Use the books together as resources to make your job or your business more productive.

Thumb through the pages here. Notice the tip boxes. You will find that some key components of the code are highlighted in the tip boxes. Go ahead, take a peek. While you are at it, pick a few topics that you are either familiar with or confused about, and look them up. See for yourself how easy this book makes it to put your mind at ease on code issues.

You might find another book that attempts to do what this one does, but if you are looking for one guide to the code that you can trust, this is it.

Acknowledgments

I would like to thank the International Code Council, Inc. for its cooperation and permission to reprint certain illustrations and tables.

2009 INTERNATIONAL PLUMBING CODE NEED TO KNOW

CHAPTER

1

Definitions

Definitions are not exactly exciting reading, but they can play a vital role in the plumbing code. It would be easy to justify skipping this chapter, but I urge you not to. I've been plumbing since the mid-1970s, hold a master's license, and have owned my own plumbing company since 1979. In addition to fieldwork and running my business, I've taught code classes and apprenticeship classes at Central Maine Technical Center. During all these years I've seen countless plumbers who could not give the proper definition of a term. For example, are you sure what the difference is between a stack vent and a vent stack? If you even thought of hesitating on this question, don't skip this chapter.

There are many cases of industry slang that conflict with proper definitions. Local sayings can be fine for getting the job done, but they don't cut it on license testing, and they can make it difficult to communicate with suppliers. For example, most plumbers in my region have a pet name for trap adapters. People in this area know what they are talking about, but if they move to another location, not knowing that the part they are requesting is technically a trap adapter could be a problem.

If you are in charge of permit acquisitions, takeoffs, and similar tasks where using the right word or term can be crucial, you must be up to speed on the definitions as set forth by the plumbing codes. We will use this chapter to learn and understand the correct terms and definitions. Don't feel that you have to memorize them, but become familiar enough to be comfortable when interpreting your local code book.

WORDS, TERMS, AND DEFINITIONS

ABS: Acrylonitrile-butadiene-styrene

ACCEPTED ENGINEERING PRACTICE: Any practice that conforms to accepted principles, tests, or standards. The accepted principles, tests, or standards must be approved by technical or scientific authorities.

ACCESS: Fixtures, appliances, and equipment that require access may be governed by one of two means of access. In essence, "access" refers to some means of making devices reachable. The means of access can be considered accessible (if the removal of a panel or plate is required for access) or ready (if a device can be reached immediately without the removal of a concealment device).

ACCESS COVER: An access cover is a device used to conceal plumbing that is required to be accessible. It is common for access covers to be secured with screws or bolts that can be removed easily.

ACCESSIBLE: When a device is deemed accessible, it is within code requirements for it to be concealed by a removable panel or plate. This is not to be confused with a device that is required to be readily accessible, in which case the removal of a concealment device is not allowed.

ADAPTER FITTING: Any approved fitting that can be used to connect pipes and fittings that would not otherwise fit together is an adapter fitting.

ADMINISTRATIVE AUTHORITY: A broad range of people and organizations act as an administrative authority. For example, your local plumbing inspector can be considered an administrative authority. In addition to individuals, code boards, code departments, and code agencies can be considered administrative authorities. An authorized representative of an administrative authority is also an administrative authority.

AIR-ADMITTANCE VALVE: A one-way valve designed to allow air to enter a plumbing drainage system is an air-admittance valve. The valve closes automatically by gravity and seals the vent terminal at zero differential pressure and under positive internal pressures. Many field plumbers refer to these devices as mechanical vents. Air-admittance valves are intended to allow air to enter a drainage system without the need for a vent that extends to open air through a roof or wall. Another feature of an air-admittance device is the prevention of sewer gas from entering a building. It is common for these valves to be installed under sinks and lavatories during remodeling work.

AIR BREAK: Don't confuse an air break with an air gap. Both can be found in drainage systems, but they are not the same. An air break is a piping arrangement in which a drain from a fixture or device discharges indirectly into another fixture, as in the case of a clothes washer discharging into a laundry sink. The indirect discharge must be made above the trap seal and below the flood-level rim.

AIR GAP (DRAINAGE): There are air gaps in both drainage and water-distribution systems. When dealing with a drainage system, an air gap is the unobstructed vertical distance through the open air between the outlet of a waste pipe and the flood-level rim of the receptacle receiving the discharge. An example of this would be a condensate pipe that terminates above a floor drain. The distance from the discharge pipe to the floor drain would be considered the air gap.

AIR GAP (WATER DISTRIBUTION): An unobstructed vertical distance through open air between the lowest opening from any pipe or faucet supplying water to a tank, plumbing fixture, or other device and the flood-level rim of a receptacle is considered to be an air gap.

ALTERNATIVE ENGINEERED DESIGN: It is possible for engineers and others to create plumbing systems that perform in accordance with the intent of the plumbing code, even though the system may not be piped in direct relation to the code. When this is done, the system is known as an alternative engineered design. So long as the system protects public health, safety, and welfare, it can be approved by a local administrative authority.

ANCHORS: *See* Supports.

ANTISIPHON: Devices designed to prevent siphonage are called antisiphon devices.

APPROVED: Anything meeting the required standards of an administrative authority can be considered approved.

APPROVED TESTING AGENCY: Groups or organizations established primarily to implement testing for conformance to approved standards required by an administrative authority are known as approved testing agencies or approved agencies.

AREA DRAIN: A device installed to collect storm or surface water from an open area, such as an areaway, is called an area drain.

ASPIRATOR: A device supplied with water or another fluid under positive pressure that passes through an integral orifice or construction that causes a vacuum is an aspirator. It is not uncommon for aspirators to be called suction devices.

BACKFLOW: Whenever water, other liquids, mixtures, or other substances flow into a potable water system from a source not intended to mix with the potable system, the act is known as backflow. For example, a water hose that has a fertilizer watering device attached to it and that is connected to a hose bibb could create a potentially deadly backflow if the contents of the watering device were to be sucked into the potable water system. The simple installation of a backflow preventer can avoid such disasters.

BACKFLOW CONNECTION: Any type of plumbing connection that is not protected from backflow can be considered a backflow connection.

BACKFLOW PREVENTER: Any device designed to prevent backflow into a potable water system is a backflow preventer.

BACKPRESSURE: A potential backflow problem can exist when backpressure occurs. Any pressure created in a water-distribution system this is in excess of the pressure in the system itself can cause backflow. The pressure causing this risk is known as backpressure.

BACKPRESSURE, LOW-HEAD: Pressure that is less than, or equal to, 4.33 pounds per square inch (psi) or the pressure exerted by a 10-foot column of water.

BACKSIPHONAGE: If there is contaminated backflow into a potable water system, it is called backsiphonage. This can occur when the pressure in a potable water system falls below the atmospheric pressure of the plumbing fixtures or devices.

BACKWATER VALVE: Some sewers and building drains are subject to backflow. When this is the case, backwater valves are installed to prevent drainage and waste from backing up into the building drain or sewer. You can think of a backwater valve as something of a check valve. Drainage and waste can flow out of the pipe in the proper direction but cannot back up into the pipe beyond the backwater valve.

BALLCOCK: Ballcocks are most often found in toilet tanks. A water-supply valve that is operated by means of a float and used to fill a tank with water is a ballcock. Modern versions of ballcocks are equipped with antisiphon devices that prevent water in a toilet tank from being sucked back into the potable-water supply system.

BASE FLOOD ELEVATION: Reference points are established that determine the peak elevation of a potential flood based on the likelihood of a flood within a 100-year period. Such reference points are known as base flood elevations. The base point takes into consideration the wave height of any flooding that may occur. All base-flood elevation points are established within the guidelines of local building-code requirements.

BATHROOM: Any room equipped with a bathing unit, such as a bathtub or shower, is considered to be a bathroom.

BATHROOM GROUP: A bathroom group consists of any group of plumbing fixtures that may or may not include a bidet or an emergency floor drain and that does include a water closet, a lavatory, and a bathing unit, such as a bathtub or shower. All fixtures in a bathroom group are located together on the same floor level.

BATTERY OF FIXTURES: Whenever there are two or more similar fixtures installed side by side that discharge into a common horizontal waste or soil branch, a battery of fixtures is created. One common example of this would be a battery of urinals on the wall of a public restroom.

BEDPAN STEAMER: Any fixture used to scald bedpans or urinals by the direct application of steam or boiling water is know as a bedpan steamer or a bedpan boiler--the two names are interchangeable.

BEDPAN WASHER AND STERILIZER: Plumbing fixtures installed for the purpose of washing bedpans and in which the contents of the fixture are allowed into the sanitary drainage system are called bedpan washers. Also included in this classification are fixtures that drain into the sanitary drainage system, but also provide for disinfecting utensils by scalding them with steam or hot water.

BEDPAN WASHER HOSE: Devices that are installed adjacent to a water closet or clinical sink and are supplied with hot and cold water for the purpose of cleaning bedpans are called bedpan washer hoses.

BOILER BLOWOFF: The emptying of discharge or sediment from a boiler is done through a boiler blowoff.

BRANCH: Any part of a piping system that is not a riser, a main, or a stack.

BRANCH, FIXTURE: *See* Fixture branch.

BRANCH, HORIZONTAL: *See* Horizontal branch.

BRANCH INTERVAL: The distance along a soil or waste stack that corresponds to the story height of a building (but not less than 8 feet) within which the horizontal branches from one floor or story of a building are connected to a stack.

BRANCH VENT: Any vent that connects one or more individual vents with a vent stack or a stack vent.

BRAZED JOINT: A joint made by the joining of metal parts with alloys that melt at temperatures higher than 840°F but lower than the melting temperature of the parts being joined.

BUILDING: A structure that is occupied or intended for occupancy by people.

BUILDING DRAIN: A building drain is the lowest drainage piping that receives the discharge from soil, waste, and other drainage pipes inside a building. Building drains extend to a length of up to 30 inches beyond the walls of a structure and convey their contents into what becomes the building sewer once the pipe is extended beyond the 30-inch limit.

BUILDING DRAIN, COMBINED: Some jurisdictions allow a building drain to convey both sewage and storm water in a single pipe. When this is the case, the pipe is known as a combined building drain.

BUILDING DRAIN, SANITARY: A building drain that conveys only sewage.

BUILDING DRAIN, STORM: A building drain that conveys storm water or other drainage but not sewage.

BUILDING SEWER: A pipe that begins about 30 inches away from a building and conveys sewage from the building to a public sewer or a private sewage-disposal system.

BUILDING SEWER, COMBINED: One that conveys both sewage and storm water.

BUILDING SEWER, SANITARY: One that conveys only sewage.

BUILDING SEWER, STORM: One that conveys storm water or other drainage but not sewage.

BUILDING SUBDRAIN: Any portion of a drainage system that does not drain by gravity into a building sewer.

BUILDING SUPPLY: A pipe that supplies water to a building from a water meter or other water source. Building supplies are often referred to as water services.

BUILDING TRAP: A device installed in the building drain or building sewer to prevent the circulation of air between the drainage system of the building and the building sewer. These devices are no longer common.

CERTIFIED BACKFLOW ASSEMBLY TESTER: Someone who is approved to test and maintain backflow assemblies to the satisfaction of the administrative authority.

CESSPOOL: A lined excavation in the earth that collects the discharge from drainage systems and retains organic matter while allowing liquids to seep through the bottom and sides of the lining to be absorbed in the ground.

CHEMICAL WASTE: *See* Special wastes.

CIRCUIT VENT: Any vent connecting to a horizontal drainage branch that vents two to a maximum of eight traps or trapped fixtures connected in a battery of fixtures.

CISTERN: A storage tank that is normally used to collect and store storm water for uses not associated with potable water.

CLEANOUT: An opening in a drainage system that allows access for the removal of obstructions in the piping. The most common type of cleanout is a removable plug or cap, but a removable fixture, such as a water closet, can also be considered a cleanout, as can a removable trap on a plumbing fixture, such as a sink.

CLINIC SINK: Sinks with a flush rim, an integral trap with a visible trap seal, and the basic flushing and cleansing characteristics of a water closet that are intended to receive the discharge from bedpans are known as clinic sinks.

CODE: Regulations set forth and adopted by local jurisdictions to dictate proper plumbing procedures as enforced by the administrative authority.

CODE OFFICIAL: An individual authorized to enforce the local code requirements.

COMBINATION FIXTURE: A sink or laundry tray that has two or three compartments in a single unit or a fixture that combines a sink with a laundry tray.

COMBINATION THERMOSTATIC/PRESSURE-BALANCING VALVE: A mixing valve able to sense water temperature from the outlet location and maintain it by regulating the temperature of incoming hot and cold water.

COMBINATION WASTE AND VENT SYSTEMS: A system designed to accept the drainage of sinks and floor drains without the standard use of vertical vents. These systems utilize horizontal wet venting in oversized pipes as an alternative to routine vertical venting.

COMBUSTIBLE CONSTRUCTION: Any structure containing building materials that will ignite and burn at a temperature of 1392°F or less.

COMMON: A word used to describe any part of a plumbing system that is meant to serve more than one fixture, building, system, or appliance.

COMMON VENT: A vent that serves more than one fixture and that is connected at the junction of the fixture drains or to a fixture branch that is serving the fixtures.

CONCEALED FOULING SURFACE: Any surface of a plumbing fixture that is not readily visible and that is not scoured or cleansed with each operation of the fixture.

CONDUCTOR: Storm-water piping found inside a building that conveys storm water from a roof to a storm drain, combined building sewer, or some other approved location.

CONFINED SPACE: An area that has a volume less than 50 cubic feet per 1,000 Btu/h of the aggregate input rating of all fuel-burning appliances installed in the area.

CONSTRUCTION DOCUMENTS: Materials that typically consist of graphics, blueprints, specifications, descriptions, and other requirements needed to obtain a building permit. It is expected that the documents will be drawn to scale when relevant.

CONTAMINATION: Any impairment of water quality in a potable water system that may cause public heath problems, such as through poisoning or the spread of disease.

CONTINUOUS VENT: Any vertical vent that is a continuation of the drain for which it serves as the vent.

CONTINUOUS WASTE: A piping arrangement that connects the drains from multiple fixtures to a common trap, as in the case of a double-bowl kitchen sink connected to a single trap.

CPVC: Chlorinated poly vinyl chloride.

CRITICAL LEVEL: A measurement that is used to establish the minimum height at which a backflow preventer or vacuum breaker can be installed above the flood-level rim of a fixture or receptor. Any space below the critical level is assumed to present a risk of backflow. When obvious markings are not evident to establish a critical level, the bottom of the device is considered to be the critical level.

CROSS CONNECTION: Any connection or arrangement that allows the possibility of contamination of a potable water system.

DEAD END: Any branch of a soil, waste, or vent pipe or building drain or sewer that extends for a length of 2 feet or more and ends with a plug, cap, or other closed fitting.

DEPARTMENT HAVING JURISDICTION: Any agency or organization, including but not limited to the administrative authority, that has the authority to interpret and enforce the plumbing code.

DEPTH OF WATER SEAL: A measurement of liquid, usually water, that would have to be removed from a trap to allow air to pass through.

DESIGN FLOOD ELEVATION: An elevation that is determined by using data specified on a community's legally designated flood-hazard map to identify a flood level that includes wave height.

DEVELOPED LENGTH: The full length of a section of piping, including fittings, when measured along the centerline of the pipe and fittings.

DIAMETER: Except where otherwise stated, diameter is considered to be the nominal diameter as designated commercially.

DISCHARGE PIPE: Any pipe that conveys the discharge from plumbing fixtures and/or appliances.

DOMESTIC SEWAGE: Liquid and water-borne wastes that come from ordinary household use, do not contain industrial wastes, and can be disposed of satisfactorily. There is no need for special treatment to prepare domestic sewage for disposal. In other words, domestic sewage can be disposed of in a sanitary public sewer or private sewage-disposal system without any preliminary treatment necessary.

DOWNSPOUT: A piping arrangement that extends down the exterior of a building and that carries storm water from a roof to a building storm drain, combined building sewer, or other means of satisfactory disposal.

DRAIN: A pipe that carries waste water or water-borne wastes in a building drainage system is a drain.

DRAINAGE FITTINGS: Special fittings used in a drainage system that are recessed and tapped to eliminate ridges on the inside of installed pipe. They differ from standard cast-iron fittings in that the latter have a bell-and-spigot design that does not offer the same smooth service that is created with drainage fittings.

DRAINAGE SYSTEM: Any piping located within private or public buildings that conveys sewage, rainwater, or other liquid waste to a point of disposal. Excluded from the drainage system is any main or public sewer system or sewage treatment or disposal plant.

DRAINAGE SYSTEM, GRAVITY: A drainage system that drains entirely by gravity to a building sewer.

DRAINAGE SYSTEM, SANITARY: A drainage system that carries sewage but not storm, surface, or ground water.

DRAINAGE SYSTEM, STORM: A drainage system that conveys storm water, surface water, condensate waste, and other similar liquids.

DURHAM SYSTEM: A special type of soil or waste drainage system that is created with threaded pipe or tubing and that utilizes recessed drainage fittings.

EFFECTIVE OPENING: This term can have multiple meanings. Generally, it refers to the minimum cross-sectional area at a point where water-supply discharge is measured or expressed in terms of the diameter of a circle. In cases where the opening is not circular in shape, the diameter measurement of a circle of equivalent cross-sectional area is used. In addition to these conditions, "effective opening" can also apply to an air gap.

EMERGENCY FLOOR DRAIN: A floor drain that does not receive the discharge of any drain or indirect waste pipe and that protects against damage from accidental spills, fixture overflows, and leakage.

ESSENTIALLY NONTOXIC TRANSFER FLUIDS: There are many types of fluids that can be considered essentially nontoxic when they have a Gosselin rating of 1. The most common fluids of this type include propylene glycol, polydimethylsiloxane, mineral oil, hydrochlorofluorocarbon, chlorofluorocarbon, and hydrofluorocarbon refrigerants, as well as FDA-approved boiler water additives for steam boilers.

ESSENTIALLY TOXIC TRANSFER FLUIDS: Any soil, waste, or gray water or fluids that have a Gosselin rating of 2 or more are considered to be essentially toxic transfer fluids. These fluids can include ethylene glycol, hydrocarbon oils, ammonia refrigerants, and hydrazine.

EXISTING INSTALLATIONS: Plumbing work and systems that were installed prior to the effective date of the current plumbing code, for which a permit was issued, are considered to be existing installations.

EXISTING WORK: *See* Existing installations.

FAUCET: A device attached to the end of a water-supply pipe that makes it possible to draw water that is being held in the pipe.

FILL VALVE: A water valve that supplies water to a device, such as a water closet. The device is opened or closed by means of a float or similar device. Antisiphon devices are an integral part of a fill valve. The antisiphon device is positioned on the discharge side of a water-supply control valve.

FIXTURE: *See* Plumbing fixture.

FIXTURE BRANCH: This term can have two meanings. A fixture branch can be a drain that serves two or more fixtures and discharges into another drain or stack. When related to a water-supply system, a fixture branch is a water-supply pipe that runs from a water distribution pipe to a fixture.

FIXTURE DRAIN: A section of drainpipe that runs from the trap of a fixture to a junction with another drainpipe.

FIXTURE FITTING: A fixture fitting is a fitting that controls the volume and/or directional flow of water to a fixture. Fixture fittings are generally attached to a fixture but may simply be accessible from the fixture. A waste fitting is a combination of components that conveys the sanitary waste from the outlet of a fixture to the connection to the sanitary drainage system.

FIXTURE SUPPLY: A pipe or tube that connects a fixture to a branch water supply or to a main water-supply pipe.

FLAMMABLE VAPOR OR FUMES: A concentration of flammable constituents in air that exceeds 25 percent of its lower flammability limit.

FLOODED: A condition that occurs when a fixture is filled with liquid that rises to the flood-level rim.

FLOOD-LEVEL RIM: The upper edge of a fixture where water will overflow if its height is greater than the edge of the fixture.

FLOOD HAZARD AREA: A flood hazard area or zone is established by choosing the greater of one of two areas. The two considerations are the area within a floodplain subject to a 1 percent or greater chance of flooding in a given year and an area that is designated as a flood hazard area on a community's flood hazard map or as otherwise legally designated.

FLOW PRESSURE: A measurement of water pressure in a pipe that is near a faucet or water outlet. The flow pressure is established when the faucet or water outlet is in a full-open position.

FLUSH TANK: A tank that is usually controlled by a ballcock and equipped with a flush valve that holds water that is released on demand to flush the contents of a bowl or other portion of a fixture, as in the case of water closets and urinals.

FLUSH VALVE: A valve located at the base of a flush tank that provides for the flushing of water closets and similar fixtures.

FLUSHOMETER TANK: A device designed to be installed within an air accumulator vessel for the purpose of discharging a predetermined quantity of water to fixtures for flushing purposes.

FLUSHOMETER VALVE: A valve that provides a predetermined quantity of water to fixtures for flushing purposes and is actuated by direct water pressure. The valve gradually closes to reseal fixture traps and to avoid water hammer.

GANG OR GROUP SHOWER: Two or more showers in a common area.

GRADE: An amount of slope or fall of a pipe in reference to a horizontal plane. Grade is also frequently called pitch. While grade or pitch can be a factor in various types of piping, it is most commonly encountered when working with drainage systems.

GREASE I\INTERCEPTOR: A passive interceptor device whose rated flow exceeds 50 gallons per minute (GPM).

GREASE-LADEN WASTE: An effluent discharge that is a byproduct of food processing, food preparation, or other source in which grease, fats, and oils enter an automatic-dishwasher prerinse station, sink, or other appurtenance.

GREASE TRAP: A passive interceptor device whose rated flow is 50 GPM or less.

HANGERS: *See* Supports.

High hazard: *See* Contamination.

HORIZONTAL BRANCH: Any pipe that extends laterally from a soil or waste stack or building drain, with or without vertical sections or branches, receives the discharge from one or more fixture drains, and conducts its contents to a soil or waste stack or to a building drain.

HORIZONTAL PIPE: A pipe or fitting that makes an angle of less than 45 degrees with the horizontal.

HOT WATER: Water with a temperature equal to or greater than 120°F.

HOUSE DRAIN: *See* Building drain.

HOUSE SEWER: *See* Building sewer.

HOUSE TRAP: *See* Building trap.

INDIRECT WASTE PIPE: A pipe that discharges into a drainage system through an air break or air gap without attaching directly to the drainage piping.

INDIVIDUAL SEWAGE-DISPOSAL SYSTEM: Any approved system that uses a septic tank, cesspool, or mechanical treatment to dispose of domestic sewage in a way that does not rely on a public sewer or public treatment facility.

INDIVIDUAL VENT: A single vent that vents a fixture trap and that either connects to a vent system above the fixture being served or terminates into open air.

INDIVIDUAL WATER SUPPLY: Any water supply, other than an approved public water supply, that serves one or more families.

INDUSTRIAL WASTE: All liquid or water-borne waste from industrial or commercial processes, except domestic sewage.

INTERCEPTOR: A device that separates and retains deleterious, hazardous, or undesirable matter from normal waste. Interceptors may be operated automatically or manually and must allow normal waste and sewage to pass through the device.

INVERT: The lowest portion of the inside of a horizontal pipe.

JOINT, BRAZED: A joint made by joining metal parts with alloys that melt at temperatures higher than 840°F but lower than the melting temperature of the parts being joined.

JOINT, EXPANSION: A piping arrangement that allows for the expansion and contraction of the piping system. Loops, return bends, and return offsets are used to create expansion joints. The primary need for this type of arrangement is found where there may be a rapid change of temperature, such as in power plants and steam rooms.

JOINT, FLEXIBLE: A type of joint between two pipes that will allow one pipe to be moved without moving the other pipe.

JOINT, MECHANICAL: *See* Mechanical joint.

JOINT, SLIP: A joint that is made by means of a washer or a special type of packing compound in which one pipe is slipped into the end of an adjacent pipe.

JOINT, SOLDERED: A joint obtained by joining metal parts with metallic mixtures or alloys that melt at a temperature up to and including 840°F.

LABELED: Materials, fixtures, equipment, and devices bearing the label of an approved agency.

LAVATORIES IN SETS: Two or three lavatories that are served by a single trap.

LEAD-FREE PIPE AND FITTINGS: Pipes and fittings containing no more than 8.0 percent lead.

LEAD-FREE SOLDER AND FLUX: Solder and flux containing no more than 0.2 percent lead.

LEADER: Exterior drainage pipe that conveys storm water from a roof or gutter drain to an approved means of disposal.

LIQUID WASTE: Any discharge from a fixture, appliance, or appurtenance in connection with a plumbing system that does not receive fecal matter.

LISTED: *See* Labeled.

LISTING AGENCY: Any agency approved by the administrative authority that is responsible for the listing and/or labeling of materials and the ongoing inspection, testing, and reporting of the materials.

LOCAL VENT STACK: A type of vent used in connection with bedpan washers. The vent is a vertical pipe to which connections are made from the fixture side of traps and through which vapor or foul air is removed from the fixtures of devices being vented.

LOT: A single or individual parcel or area of land that is legally recorded or validated by any means acceptable to the administrative authority on which a building is situated or is the site of any work regulated by the code. This includes yards, courts, and unoccupied spaces legally required for the building or works and owned by or in the lawful possession of the owner of the building or works.

LOW HAZARD: *See* Pollution.

MACERATING TOILET SYSTEMS: An assembly that consists of a water closet and sump with a macerating pump that is designed to collect, grind, and pump wastes from the water closet and up to two other fixtures connected to the sump.

MAIN: A principal pipe artery to which branches are connected.

MAIN SEWER: *See* Public sewer.

MAIN VENT: A principal pipe artery of a vent system to which the vent branches may be connected.

MANIFOLD: *See* Plumbing appurtenance.

MAY: A permissive term.

MECHANICAL JOINT: Typically, a joint that is made by applying compression along the centerline of the pieces being joined. The joint may be part of a coupling, fitting, or adapter. Mechanical joints are not screwed, caulked, threaded, soldered, solvent-cemented, brazed, or welded.

MEDICAL GAS SYSTEM: A complete system used to deliver medical gases for direct patient application from a central supply system through piping networks with pressure and operating controls, alarm-warning systems, and related components,and extending to station outlet valves at patient use points.

MEDICAL VACUUM SYSTEM: A system consisting of central vacuum-producing equipment with pressure and operating controls, shutoff valves, alarm-warning systems, gauges, and a network of piping extending to and terminating with suitable station inlets at locations where patient suction may be required.

MOBILE-HOME-PARK SEWER: Part of a horizontal piping drainage system that begins 2 feet downstream from the last mobile-home site and conveys it to a public sewer, private sewer, individual sewage-disposal system, or some other point of disposal.

NONPOTABLE WATER: Water that is not safe for drinking or personal or culinary use.

NUISANCE: A nuisance can be any inadequate or unsafe water-supply or sewage-disposal system. If work regulated by the code is done in such a way as to be dangerous to human life or detrimental to health and property, the act is a nuisance. Any public nuisance known in common law or equity jurisprudence is also a nuisance in this sense.

OCCUPANCY: The purpose for which a building or portion thereof is utilized or occupied.

OFFSET: Any combination of elbows or bends in a line of piping that brings one section of the pipe out of line but into a line parallel with the other section.

OIL INTERCEPTOR: *See* Interceptor.

OPEN AIR: Fresh air outside a structure.

PB: Polybutylene.

PE: Polyethylene.

PERSON: A natural person, an heir, an executor, administrator, or assign, also including a firm, corporation, municipal or quasi-municipal corporation, or governmental agency. Singular includes plural and male includes female.

PIPE: Any cylindrical conduit or conductor conforming to the particular dimensions commonly known as pipe size.

PLUMBING: Any business, trade, or work that has to do with the installation, removal, alteration, maintenance, or repair of plumbing and drainage systems or parts thereof. This work relates to the connection of sanitary drainage systems, storm drainage systems, venting systems, and public or private water-supply systems.

PLUMBING APPLIANCE: A special class of plumbing fixtures that is meant to perform special functions. Fixtures that depend upon motors, controls, heating elements, pressure-sensing, or temperature-sensing elements can all be appliances.

PLUMBING APPURTENANCE: A device that is an adjunct to a basic piping system and plumbing fixtures. Appurtenances do not demand any additional water supply and do not add any discharge load to a fixture or drainage system.

PLUMBING FIXTURE: A receptacle or device that is either permanently or temporarily connected to a water-distribution system of the premises and demands a supply of water therefrom, discharges waste water, liquid-borne waste materials, or sewage either directly or indirectly to the drainage system of the premises, or requires both a water-supply connection and a discharge to the drainage system of the premises.

PLUMBING OFFICIAL: *See* Administrative authority.

PLUMBING SYSTEM: Includes a water-supply and distribution pipes, plumbing fixtures, traps, water-treating equipment, water-using equipment, soil pipes, waste pipes, vent pipes, sanitary sewers, storm sewers, and building drains. The system can also include connections, devices, and appurtenances within a structure or premises.

POLLUTION: Impairment of the quality of potable water in an amount sufficient to cause disease or harmful physiological effects and conforming in bacteriological and chemical quality to the requirements of the Public Health Service Drinking Water Standards or the regulations of the public health authority having jurisdiction.

POTABLE WATER: Water that is safe and suitable for drinking, culinary purposes, and domestic purposes.

PP: Polypropylene.

PRESSURE: Typical amount of force exerted by a homogeneous liquid or gas, per unit of area, on the wall of a container or conduit.

PRESSURE, RESIDUAL: The usable amount of water pressure available at a fixture or water outlet after allowances have been made for pressure drops caused by friction loss, head, and other reasons for decreased pressure.

PRESSURE, STATIC: The amount of pressure present when there is no flow.

PRESSURE-BALANCING VALVE: A mixing valve that receives both hot and cold water and keeps the pressure stable by compensating for fluctuations in either hot or cold water.

PRIVATE: As used in the plumbing code, this refers to plumbing fixtures that are not intended for use by the general public. Fixtures installed in residences, rooms of hotels and motels, and other facilities intended for use by either a family or an individual are considered private.

PRIVATE SEWAGE-DISPOSAL SYSTEM: Most often consists of a septic tank that allows effluent to discharge into a subsurface septic field. However, any sewage-disposal system that meets code criteria and that does not discharge into a public sewer can be considered a private system.

PRIVATE SEWER: Any pipe that receives drainage from more than one building drain and then conveys the drainage to a public sewer or private sewage-disposal system.

PUBLIC SEWER: A common sewer that is controlled by any public authority.

PUBLIC USE: Any use that is not defined as private.

PUBLIC WATER MAIN: A primary water-supply pipe that is controlled by any public authority.

PVC: Polyvinyl chloride.

QUICK-CLOSING VALVE: Any valve or faucet that closes automatically when released manually or that is controlled by a mechanical means for closing quickly.

READY ACCESS: Means of direct access to a fixture or device. To qualify as ready access the access must be possible without any need for the removal of a panel, the opening of a door, or any other obstruction. Additionally, access must be possible without the need of a ladder or other similar device.

RECEPTOR: An approved fixture or device that is used to accept the discharge from indirect waste pipes and is able to be cleaned readily.

REDUCED PRESSURE PRINCIPLE BACKFLOW PREVENTER: Any backflow preventer that contains two independently acting check valves. The check valves are internally force-loaded to a normally closed position and separated by an intermediate chamber in which there is an automatic relief means of venting to atmosphere, internally loaded to a normally open position between two tightly closing shutoff valves, and with means for testing for tightness of the checks and opening of relief means.

REGISTERED DESIGN PROFESSIONAL: An architect or engineer who is registered or licensed to practice his or her profession within the guidelines of the governing agency.

REGULATING EQUIPMENT: Any valve or control used in a plumbing system that is required to be either accessible or readily accessible.

RELIEF VALVE, PRESSURE: A valve that is pressure-actuated and held closed by a spring or other means that serves to relieve pressure automatically when a set pressure is reached.

RELIEF VALVE, TEMPERATURE: A relief valve that opens when a set temperature is reached.

RELIEF VALVE, TEMPERATURE ANDPRESSURE: A relief valve that opens when a set pressure or a set temperature is reached.

RELIEF VENT: Any vent that provides air circulation between drainage and vent systems.

REMOTE OUTLET: When used for sizing water piping, a remote outlet is the furthest outlet dimension, measuring from the meter, either the developed length of the cold-water piping or through the water heater to the furthest outlet of the hot-water piping.

RIM: The unobstructed open edge of a fixture.

RISER: Any water-supply pipe that extends vertically for one full story or more to convey water to branches or fixtures.

ROOF DRAIN: Any drain that is installed to receive water from a roof and then convey the water to a suitable discharge location.

ROUGH-IN: Any part of a plumbing system that is installed prior to the installation of plumbing fixtures.

SAND INTERCEPTOR: *See* Interceptor.

SELF-CLOSING FAUCET: Any faucet that closes automatically once deactivation of the opening means is created.

SDR: Standard dimensional ratio.

SEEPAGE PIT: Any lined excavation in the ground that accepts discharge from a septic tank and then allows the effluent to seep into the earth from the bottom and sides of the seepage pit.

SEPARATOR: *See* Interceptor.

SEPTIC TANK: Any approved container, usually made of concrete, that is buried in the ground and accepts the discharge from a drainage system or pipe. Septic tanks must be watertight and designed to retain solids and digest organic matter via a period of detention and to allow effluent to flow into a septic field or other approved discharge destination.

SEWAGE: Liquid waste that contains animal or vegetable matter in suspension or solution and may include liquids containing chemicals in solution.

SEWAGE EJECTOR: A device that lifts or pumps sewage by entraining the sewage in a high-velocity jet of steam, air, or water.

SEWAGE PUMP: A mechanical device, other than an ejector, that is installed permanently to remove sewage or liquid waste from a sump.

SEWER, BUILDING: *See* Building sewer.

SEWER, PUBLIC: *See* Public sewer.

SEWER, SANITARY: A sewer that conveys sewage without combining storm, surface, or ground water with it.

SEWER, STORM: A sewer that conveys storm water, surface water, condensate, cooling water, and similar liquid wastes.

SHALL: A mandatory term.

SHIELDED COUPLING: Any approved elastomeric sealing gasket that is equipped with an approved outer shield and a mechanism for tightening.

SHOCK ARRESTOR: *See* Water hammer arrestor.

SINGLE-FAMILY DWELLING: A structure that is constructed for the purpose of housing the property owner and family. The structure must be the only dwelling located on a parcel of ground with typical accessory buildings but no other homes.

SIZE AND TYPE OF TUBING: *See* Diameter.

SLIP JOINT: An adjustable tubing connection that consists of a compression nut, a friction ring, and a compression washer and is designed to fit a threaded adapter fitting or standard-taper pipe thread.

SLOPE: *See* Grade.

SOIL PIPE: Any pipe that conveys sewage containing fecal matter to a building drain or building sewer.

SOLDERED JOINT: A joint created by joining metal parts with metallic mixtures or alloys that melt at a temperature below 800°F and above 300°F.

SPECIAL WASTES: Any waste that requires special handling or treatment.

SPILL-PROOF VACUUM BREAKER: An assembly of one check valve that is force-loaded closed and an air-inlet vent valve that is force-loaded open to atmosphere downstream of the check valve, located between and including two tightly closing shutoff valves and a test cock.

STACK: A vertical pipe that is part of a soil, waste, or vent system and rises to a height of at least one story.

STACK VENT: A continuation of a soil or waste stack above the highest horizontal drain connected to the stack.

STACK VENTING: Using a stack vent to vent a soil or waste stack.

STERILIZER, BOILING-TYPE: A fixture of a nonpressure type utilized for boiling devices for disinfection.

STERILIZER, INSTRUMENT: A device used to sterilize instruments.

STERILIZER, PRESSURE: A device used for sterilization that consists of a vessel using steam under pressure.

STERILIZER, PRESSURE INSTRUMENT WASHER: A pressure-vessel fixture that washes and sterilizes instruments.

STERILIZER, UTENSIL: A sterilizer used to sterilize utensils.

STERILIZER VENT: A pipe or stack that is connected indirectly to a drainage system at a lower terminal to receive the vapors from nonpressure sterilizers or the exhaust vapors from pressure sterilizers. The vent then transports the vapors directly to open air. Other names for a sterilizer vent can include vapor, steam, atmospheric, or exhaust vent.

STERILIZER, WATER: A device used to sterilize and then store sterilized water.

STORM DRAIN: *See* Building drain, storm.

STORM SEWER: *See* Sewer, storm.

STRUCTURE: Anything that is built or constructed or any portion thereof.

SUBSOIL DRAIN: A drain used to collect subsurface or seepage water and convey it to an approved disposal location.

SUMP: A container or pit located below the normal grade of a gravity system that is used to accept sewage or liquid waste that will be pumped out of the holding device.

SUMP PUMP: An electric pump that works automatically to remove the contents of a sump that does not contain raw sewage.

SUMP VENT: A vent from a pneumatic sewage ejector or similar device that extends to open air.

SUPPORTS: Devices used to support and or secure pipes, fixtures, and equipment.

SWIMMING POOL: Any structure, container, or other device that contains an artificial body of water for the purposes of swimming, diving, or recreational bathing and that has a depth of 2 feet or more at any point.

TAILPIECE: A pipe or tube that connects the outlet of a plumbing fixture to a trap.

TEMPERED WATER: Water with a temperature range from 85 to 120°F.

THERMOSTATIC VALVE: Also known as a temperature-control valve, this valve is designed to mix hot and cold water while compensating for temperature fluctuations to maintain an even water temperature at the point of delivery.

TRAP: A fitting or device that holds water to prevent the emission of sewer gas without materially affecting the flow of sewage or waste water through a pipe.

TRAP ARM: A section of pipe that extends from a fixture drain's trap to a drain.

TRAP PRIMER: A device or system used to maintain a suitable amount of water in a trap that sees little use and that would potentially dry up without the aid of the primer.

TRAP SEAL: The vertical distance between the weir and the top of the dip of a trap.

TYPE A DWELLING UNIT: Any dwelling unit designed and built for accessibility in accordance with the provisions of CABO/ANSI A117.1.

TYPE B DWELLING UNIT: Any dwelling unit designed and built in accordance with the provisions of CABO/ANSI A117.1.

UNCONFINED SPACE: Any room, space, or area that has a volume equal to at least 50 cubic feet per 1000 Btu/h of the aggregate input rating of all fuel-burning appliances installed in the room, space, or area.

UNSANITARY: A condition that creates a risk to public health and sanitary principles.

UNSTABLE GROUND: Earth that does not provide a uniform bearing for the barrel of a sewer pipe between the joints at the bottom of the pipe trench.

VACUUM: Pressure that is less than that exerted by the atmosphere.

VACUUM BREAKER: *See* Backflow preventer.

VACUUM RELIEF VENT: A device that doesn't allow excessive pressure to develop in a pressure vessel.

VENT: A pipe used to ventilate a plumbing system to prevent trap siphonage and backpressure or to equalize the air pressure within the drainage system.

VENT PIPE: *See* Vent.

VENT STACK: A vertical pipe that is installed to provide air circulation to a drainage system.

VENT SYSTEM: A pipe or system of piping used to ventilate a plumbing system to prevent trap siphonage and backpressure or to equalize the air pressure within the drainage system.

VENTED APPLIANCE CATEGORIES: Category I is an appliance that operates with a nonpositive vent static pressure and with a vent gas temperature that avoids excessive condensate production in the vent. Category II is an appliance that operates with a nonpositive vent static pressure and with a vent gas temperature that may cause excessive condensate production in the vent. Category III is an appliance that operates with a positive vent static pressure and with a vent gas temperature that avoids excessive condensate production in the vent. Category IV is an appliance that operates with a positive vent static pressure and with a vent gas temperature that may cause excessive condensate production in the vent.

VERTICAL PIPE: A pipe or fitting that is installed in a vertical position or that doesn't make an angle of more than 45 degrees with the vertical.

WALL-HUNG WATER CLOSET: A water closet that is mounted on a wall so that the fixture does not touch the floor.

WASTE: Discharge from any fixture, appliance, area, or appurtenance that does not contain fecal matter.

WASTE PIPE: A pipe that conveys only waste.

WATER CONDITIONING OR TREATING DEVICE: A device that conditions or treats a water supply to change its chemical content or to remove suspended solids through filtration.

WATER-DISTRIBUTING PIPE: A pipe in a building that conveys potable water from a building supply pipe to plumbing fixtures and water outlets.

WATER HAMMER ARRESTOR: A device used to absorb pressure surge that occurs when water flow is suddenly stopped in a water-supply system.

WATER MAIN: A water-supply pipe that provides water for public or community usage.

WATER OUTLET: Any discharge opening through which water is supplied to a fixture or into the atmosphere, except into an open tank that is part of a water-supply system, into a boiler or heating system, or into any devices or equipment requiring water to operate but not part of a plumbing system.

WATER PIPE, RISER: Any water-supply pipe that rises at least one full story to convey water to branches or a group of fixtures.

WATER PIPE, WATER-DISTRIBUTION: A pipe within a building that conveys water from a water service pipe or from a water meter when the meter is located in the building to a point of utilization.

WATER PIPE, WATER-SERVICE: A pipe from a water main or other potable water source or from a meter when the meter is at the public right of way to the water-distribution system in the building being served.

WATER-SUPPLY SYSTEM: Components used to create a water-supply system can include a water service pipe, water distribution pipes, and all needed connecting pipes, fittings, and control valves, as well as all appurtenances in or adjacent to a structure.

WELDED JOINT OR SEAM: A joint or seam that is obtained by joining metal parts in a plastic molten state.

WELDER, PIPE: A person who specializes in the welding of pipes and who holds a valid certificate of competency from a recognized testing laboratory.

WELL, BORED: Any well that is created by boring a hole in the earth with an auger and fitting with a casing.

WELL, DRILLED: Any well made with a drilling machine and fitted with a casing or screen.

WELL, DRIVEN: Any well created when a pipe is driven into the earth.

WELL, DUG: A well created by digging a large-diameter hole in the ground and installing a casing.

WET VENT: Any vent that serves as both vent and drain.

WHIRLPOOL BATHTUB: A bathtub that is fitted with a circulating piping system designed to accept, circulate, and discharge bath water when used.

YOKE VENT: A pipe that connects upward from a soil or waste stack to a vent stack for the purpose of preventing pressure changes in the stacks.

CHAPTER

2

Administration

Administrative polices and procedures are what make the plumbing code effective. Without the proper procedures and administration, the plumbing code would be little more than an organized outline for good plumbing procedures. To be effective, the code must be enforced. To be fair, the rules for the administration of the code must be clear to all who work with it. Administrative policies dictate the procedure for code enforcement, interpretation, and implementation.

The rules, regulations, and laws that comprise the plumbing code are structured around facts. These facts are the result of research into ways to protect the health of our nation. As a plumber, you are responsible for the health and sanitation of the public. A mistake or a code violation could result in widespread illness or even death.

In some jurisdictions the plumbing code is comprised of rules. In other areas it is a compilation of laws. There is a big difference between a rule and a law. When you are working with a rule-based code, you are subject to various means of punishment for violating the rules. The punishment may be the suspension or revocation of your license. There may be cash fines required for violations, but there is no jail time. In jurisdictions using a law-based code you could find yourself behind bars for violating the plumbing code.

The procedure required to obtain a plumber's license is not easy. Some people feel that there are too many restraints in the licensing requirements for plumbers, but they are not aware of the heavy responsibility plumbers must bear. The public often perceives a plumber as someone who works in sewers, has a poor education, and is slightly more than a common laborer. This is a false perception.

Professional plumbers are much more than sewer rats. Today's plumbers are generally well educated and have the ability to perform highly technical work. The mathematical demands on a plumber could perplex many people. The mechanical and physical abilities of plumbers are often outstanding.

While drain cleaning is a part of the trade, so is the installation of $2,500 gold faucets. The plumbing trade is not all sewage and water. Plumbers are known for their fabled high incomes, and it is true that good plumbers make more money than many professionals. The trade can offer a prosperous living, but it must be worked for.

Whether you are a plumber or an apprentice, you have much to be proud of. The plumbing trade is more than a job. As a professional plumber you will have the satisfaction of knowing you are helping to maintain the health of the nation and the integrity of our natural resources.

Not so long ago there was no plumbing code. People could pollute our lakes and streams with their ineffective cesspools and outhouses. The plumbing code is designed to stop pollution and health hazards. It is part of any plumber's life and career. Learning the code can be a laborious task, but the self-satisfaction obtained when you master the code is well worth the effort.

CODE UPDATE

Where there is a conflict between a general requirement and a specific requirement, the specific requirement governs.

Fast Fact

The plumbing code is intended to be interpreted by the local code-enforcement officer. The interpretation of the code officer may not be the same as yours, but it is the code officer's option to determine the meaning of the code.

When you earn your master's license, you will have the opportunity to run your own shop. Having your own plumbing business can be quite profitable. To realize these profits, you must first learn the code and the trade. Then you must pass the tests for your journeyman and master licenses. During your learning stages you are earning a good wage and providing a vital service to the community.

There are few professions that allow you to earn a good living while you are gaining the skills necessary to master your craft. When you attend college, you must pay for your education. With plumbing, you get paid to learn. After your training you are unlimited in the wealth you can build from your own business.

The first step toward financial independence as a plumber is a clear understanding of the plumbing code. Unlicensed plumbers are not allowed to work in many jurisdictions. Your license is your ticket to respectable paychecks and a solid future. Let's see how the administrative policies and procedures in the plumbing code will affect you.

CODE UPDATE

All new plumbing work, even on existing plumbing, must meet the provisions of the current code requirements. The exception to this is if the work is done in the same manner and arrangement as it was in the existing work and is not hazardous and is approved.

Fast Fact

The plumbing code requires any alterations or repairs to an existing plumbing system to conform to the general regulations of the code as they would apply to new installations.

WHAT DOES THE PLUMBING CODE INCLUDE?

The plumbing code includes all major aspects of plumbing installations and alterations. Design methods and installation procedures are a cornerstone of the plumbing code. Sanitary piping for the disposal of waste, water, and sewage is controlled by the code. Potable water supplies fall under the jurisdiction of the plumbing code. Storm water, gas piping, chilled-water piping, hot-water piping, and fire sprinklers are all dealt with in the plumbing, mechanical, and building codes. Any installation of fuel-gas-distribution piping and equipment, fuel-gas-fired water heaters, and water-heater venting systems are regulated by the International Fuel Gas Code.

The plumbing code is meant to ensure the proper design and installation of plumbing systems and to ensure public health and safety. The plumbing code is intended to be interpreted by the local code-enforcement officer. The interpretation of the code officer may not be the same as yours, but it is the code officer's option to determine the meaning of the code.

HOW THE CODE PERTAINS TO EXISTING PLUMBING

The plumbing code requires any alterations or repairs to an existing plumbing system to conform to the general regulations of the code as they

Trade Tip

Always obtain a plumbing permit and required inspections when replacing a water heater.

Trade Tip

In general, if you are only doing minor repair or maintenance work, you are not required to update the present plumbing conditions to current code requirements.

would apply to new installations. No alteration or repair should cause an existing plumbing system to become unsafe. Further, the alterations or repairs shall not be allowed to have a detrimental effect on the operation of the existing system.

For example, if a plumber is altering an existing system to add new plumbing, he or she must make all alterations in compliance with code requirements. It would be a violation of the code to add new plumbing to a system that was not sized to handle the additional load of the increased plumbing.

There are provisions in the code to allow existing conditions that are in violation of the current code to be used legally. If an existing condition was of an approved type prior to the current code requirements, that existing condition may be allowed to continue in operation so long as it is not creating a safety or health hazard.

If the use or occupancy of a structure is being changed, the change must be approved by the proper authorities. It is a violation of the plumbing code to change the use or occupancy without the proper approvals. For example, it would be a breach of the code to convert a residential dwelling to a professional building without the approval of the code-enforcement office.

CODE UPDATE

Provisions of the plumbing code are not deemed to nullify and provisions of local, state, or federal laws.

Fast Fact

The local code officer has the authority to alter the provisions of the plumbing code, so long as the health, safety, and welfare of the public are not endangered.

SMALL REPAIRS

Small repairs and minor replacements of existing plumbing may be made without bringing the entire system into compliance with the current plumbing-code standards. These changes must be made in a safe and sanitary method and must be approved.

For example, it would be permissible to repair a leak in a half-inch pipe without changing to a three-quarter-inch pipe, even if the current code required the larger size for the present use. You would also be allowed to replace a defective S-trap with a new S-trap, even though S-traps are not in compliance with the current code requirements. In general, if you are only doing minor repair or maintenance work, you are not required to update the present plumbing conditions to current code requirements.

It is incumbent upon the owner of a property to keep the plumbing system in good and safe repair. The owner may designate an agent to assume responsibility for the condition of the plumbing, but it is mandatory that the plumbing be kept safe and sanitary at all times.

RELOCATION AND DEMOLITION OF EXISTING STRUCTURES

If a building is moved to a new location, the building's plumbing must conform to the current code requirements of the jurisdiction where the structure will be located.

CODE UPDATE

A code officer can appoint deputies as code officials—technical officers, inspectors, and other employees.

CODE UPDATE

Code officers cannot be held personally responsible for actions related to the discharge of their duties.

In the event that a structure is to be demolished, it is the owner's, or the owner's designated agent's, responsibility to notify all companies, persons, and entities having utilities connected to the structure. These utilities may include, but are not limited to, water, sewer, gas, and electrical connections.

Before the building can be demolished or moved, the utilities having connections to the property must be disconnected and their connections sealed in an approved manner. This applies to water meters and sewer connections as well as other utilities.

MATERIALS

All materials used in a plumbing system must be approved for use by the code-enforcement office. These materials must be installed in accordance with the requirements of the local code authority. The local code officer has the authority to alter the provisions of the plumbing code so long as the health, safety, and welfare of the public are not endangered.

A property owner, or that owner's agent, may request a variance from the standard code requirements when conditions warrant a hardship. It is the code officer's decision as to whether or not the variance should be granted. The application for a variance and the final decision of the code officer must be in writing and filed with the code-enforcement office.

The use of previously used materials is open to the discretion of the local code officer. If the used materials have been reconditioned, tested, and are

CODE UPDATE

Code officials are responsible for keeping records of all aspects of their department and duties.

Fast Fact

Plans and specifications may not be required for the issuance of a plumbing permit. However, if plans and specs are required, a riser diagram and a general blueprint of the structure may be needed.

in working condition, the code officer may allow their use in a new plumbing system.

Alternative materials and methods not specifically identified in the plumbing code may be allowed under certain circumstances. If the alternatives are equal to the standards set forth in the code for quality, effectiveness, strength, durability, safety, and fire resistance, the code officer may approve the use of the alternative materials or methods. Where the requirements of reference standards or manufacturer's installation instructions do not conform to minimum provisions of thee code, the provision of the code will apply.

Before alternative materials or methods are allowed for use, the code officer can require adequate proof of the properties of the materials or methods. Any costs involved in testing or providing technical data to substantiate the use of alternative materials or methods is the responsibility of the permit applicant.

CODE OFFICERS

Code officers are responsible for the administration and enforcement of the plumbing code. They are appointed by the executive authority for the community. Code officers may not be held liable on a personal basis when working for a jurisdiction. Legal suits brought against code officers arising from on-the-job disputes will be defended by the legal representative for the jurisdiction.

CODE UPDATE

Code officers have the authority to waive some code requirements when a modification is needed for a specific reason.

CODE UPDATE

Code officers are authorized to inspect and evaluate systems, equipment, buildings, devices, premises, and spaces and areas to be used prior to issuing a permit.

The primary function of code officers is to enforce the code. Code officers are also responsible for answering questions pertinent to the materials and installation procedures used in plumbing. When an application is made for a plumbing permit, the code officer is the individual who receives the application. After reviewing a permit application, the code officer will issue or deny a permit.

Once a permit is issued by the code officer, it is the officer's duty to inspect all work to ensure that it is in compliance with the plumbing code. When code officers inspect a job, they are looking for more than just plumbing. These inspectors will be checking for illegal or unsafe conditions on the job site. If the safety conditions found on the site or in the plumbing installation are found to be in violation of the code, the code officer will issue a notice to the responsible party.

Code officers normally perform routine inspections in person. However, inspections may be performed by authoritative and recognized services or individuals other than the code officers. The results of all inspections must be documented in writing and certified by an approved individual.

If there is ever any doubt as to the identity of a code officer, you may request to see the inspector's identification. Code officers are required to carry official credentials while discharging their duties.

CODE UPDATE

When there is a change in occupancy a code officer must approve the proposed new use. The change must not result in any hazard to the public health, safety, or welfare.

CODE UPDATE

A permit holder who fails to allow access for a full inspection can be held responsible for the expense of making the concealed work accessible.

Another aspect of the code officer's job is the maintenance of proper records. Code officers must maintain a file of all applications, permits, certificates, inspection reports, notices, orders, and fees. These records are required to be maintained for as long as the structure they apply to is still standing, unless otherwise stated in other rules and regulations.

PLUMBING PERMITS

Most plumbing work, other than minor repairs and maintenance, requires a permit. This permit must be obtained prior to the commencement of any plumbing work. The code-enforcement office provides forms to individuals wishing to apply for plumbing permits. The application forms must be properly completed and submitted to the code-enforcement officer.

Permits are to be obtained by the person or agent who will install all or part of any plumbing system. The applicant must meet all qualifications required of a permit applicant. It is also required that the full name and address of the applicant be stated in the permit application.

The permit application must give a full description of the plumbing to be done. This description must include the number and type of plumbing fix-

Trade Tip

There are many plumbing codes in use. Each local jurisdiction generally amends an existing code to local needs. To be sure of your local code requirements, you must check with the local code-enforcement office.

tures to be installed. The location where the work will be done and the use of the structure housing the plumbing must also be disclosed.

The code officer may require a detailed set of plans and specifications for the work to be completed. Duplicate sets of the plans and specs may be required so that copies can be placed on file in the code-enforcement office. If the description of the work deviates from the plans and specifications submitted with the permit application, it may be necessary to apply for a supplementary permit.

The supplementary permit will be issued after a revised set of plans and specs has been given to the code officer and approved. The revised plans and specifications must show all changes in the plumbing that are not in keeping with the original plans and specs.

Plans and specifications may not be required for the issuance of a plumbing permit. However, if plans and specs are required, a riser diagram and a general blueprint of the structure may be needed. The riser diagram must be very detailed. The diagram must indicate pipe size, direction of flow, elevations, fixture-unit ratings for drainage piping, horizontal pipe grading, and fixture-unit ratings for the water-distribution system.

If the plumbing to be installed is an engineered system, the code officer may require details on computations, plumbing procedures, and other technical data. Any application for a permit to install new plumbing might require a site plan. The site plan must identify the locations of the water service and sewer connections. The locations of all vent stacks and their proximity to windows or other ventilation openings must be shown.

In the event that new plumbing is being installed in a structure served by a private sewage-disposal system, there are yet more details to be included in the site plan. When a private sewage system is used, the plan must show the

CODE UPDATE

A code officer can accept reports of approved inspection agencies when the agencies meet all requirements as to qualifications and reliability.

location of the system and all technical information pertaining to the proper operation of the system.

When a plumbing permit is applied for, the code officer will process the application in a timely manner. If the application is not approved, the code officer will notify the applicant, in writing, of the reasons for denial. If the applicant fails to follow through on the issuance of a permit within six months from the date of application, the permit request can be considered void. Once a permit is issued, it may not be assigned to another person or entity. Permits will not be issued until the appropriate fees are paid. Fees for plumbing permits are determined by the local jurisdiction.

Plumbing permits bear the signature of the code officer or an authorized representative. The plans submitted with a permit application will be labeled as approved plans by the code officer. One set of the plans will be retained by the code-enforcement office. A set of approved plans must be kept on the job site. The approved plans kept on the job must be available to the code officer or an authorized representative for inspection at all reasonable times.

It is possible to obtain permission to begin work on part of a plumbing system before the entire system has been approved. For example, you might be given permission to install the underground plumbing for a building before the entire plumbing system is approved. These partial permits are issued by the code officer, with no guarantee that the remainder of the work will be approved. If you proceed to install the partial plumbing, you do so at your own risk with regard to the remainder of the job not yet approved.

There are time limits involved with permits. If work is not started within six months of the date a permit is issued, the permit may become void. If work is started but then stalled or abandoned for a period of six months,

CODE UPDATE

Code officers are authorized to suspend or revoke a notice of approval issued under the provisions of this code wherever the notice is issued in error.

the permit may be rendered useless. A permit may be revoked if the code officer finds that the permit was issued based on false information. Misrepresentation in the application for a permit or on the plans submitted for review is reason for the revocation of a plumbing permit.

All work performed must be done according to the plans and specifications submitted to the code officer in the permit-application process. All work must be in compliance with the plumbing code. Code officers are required to conduct inspections of the plumbing being installed during the installation and upon completion of the installation.

MULTIPLE PLUMBING CODES

There are many plumbing codes in use. Each local jurisdiction generally amends an existing code to local needs. To be sure of your local code requirements, you must check with the local code-enforcement office. This book is written to explain good plumbing procedures. However, various jurisdictions have different opinions of what good plumbing procedures are. Some states, counties, or towns adapt an exiting code without much revision. Other areas make significant changes in the established code that is used as a model. It would not be unheard of to find a jurisdiction working with regulations from multiple plumbing codes. In light of these facts, always check with your local authorities before performing plumbing work.

CHAPTER

3

General Regulations

When working with the plumbing code, you must be aware of the requirements and procedures involved with regulations, permits, and enforcement. These three elements work together in ensuring the proper use of the plumbing code. This chapter is going to teach you about all three key elements, but it will also do much more. You are going to learn the facts about pipe protection, pipe connections, temporary toilet facilities, health, safety, and more. There will even be tips on how to work with plumbing inspectors instead of against them.

REGULATIONS

What are regulations? Regulations are the rules or laws used to control an activity, in this case the performance of plumbing. The plumbing code is governed by rules in some states and laws in others. If you violate the regulations in a rule-based state, you may face disciplinary action but not jail. In states where the plumbing regulations are laws, you risk going to jail if you violate the regulations.

Trade Tip

Generally, any existing condition that is not a hazard to health and safety is allowed to remain in an installation.

It may be difficult to imagine going to jail for violating a plumbing regulation, but it could happen. Certain violations could result in personal injury or death. To protect yourself and others, it is important to understand and abide by the regulations governing the plumbing trade.

Existing Conditions

The first regulations we are going to examine have to do with existing conditions. Many people have difficulty in determining their responsibilities for these as provided by the plumbing code. Generally, any existing condition that is not a hazard to health and safety is allowed to remain in existence. However, when existing plumbing is altered, it may have to be brought up to current code requirements.

While the code is normally based on new installations, it does apply to existing plumbing that is being altered. These alterations may include repairs, renovations, maintenance, replacements, and additions. The question of when work on existing plumbing must meet code requirements is one that plagues many plumbers. Let's clear this up.

The code is only concerned with changes being made to existing plumbing. As long as the existing plumbing is not creating a safety or health hazard and is not being altered, it does not fall under the scrutiny of most plumbing codes. If you are altering an existing system, the alterations generally must comply with the code requirements, but there may be exceptions to this rule. For example, if you are replacing a kitchen sink and there is no vent on the sink's drainage, the code would require you to vent the fixture. Where undue hardship exists in bringing an existing system into compliance, the code officer may grant a variance.

In the case of the kitchen-sink replacement, such a variance may be in the form of a permission to use a mechanical vent. Whenever you encounter a

Trade Tip

When you add new plumbing to an old system, you must be concerned with the size and ability of the old fixtures to handle the new installation. Increasing the number of fixture units entering an old pipe may force you to increase the size of the pipe.

severe hardship in making old plumbing come up to code, talk with your local code officer. He or she should be able to offer some form of assistance, either in the form of a variance or advice on how to accomplish your goal.

Since the code does come into play with repairs, maintenance, replacements, alterations, and additions, let's see how it affects each of these areas. If you are repairing a plumbing system, you must be aware of code requirements. If no health or safety hazards exist, nonconforming plumbing may be repaired to keep it in service.

Do you need to apply for a plumbing permit to replace a faucet? No, faucet replacement does not require a permit, but it does require the replacement to be made with approved materials and in an approved manner. Remember this: you do need a permit to replace a water heater. Even if the replacement heater is going in the same location and connecting to the same existing connections, you must apply for a permit and have your work inspected. An improperly installed water heater can become a serious hazard, capable of causing death and destruction. Your failure to comply with the code in these circumstances could ruin your life and the lives of others.

CODE UPDATE

Most plumbing is prohibited in elevator shafts. However, floor drains, sumps, and sump pumps are allowed when the plumbing is indirectly connected to the plumbing system.

Trade Tip

Beware of the change-in-use regulations. If you do commercial plumbing, this regulation can have a particularly serious effect on your plumbing costs and methods.

Routine maintenance of a plumbing system must be done according to the code, but it does not require a permit. Alterations to an existing system may require the issuance of a permit, depending upon the nature of the alteration. In any case, alterations must be done with approved materials and in an approved manner. When adding on to a plumbing system, you will normally need to apply for a permit. Adding new plumbing will come under the authority of the plumbing code and will generally require an inspection.

You can get yourself into deep water when adding on to an existing system. When you add new plumbing to an old system, you must be concerned with the size and capability of the old plumbing. Increasing the number of fixture units entering an old pipe may force you to increase the size of the pipe. This can be very expensive, especially when the old pipe happens to be the building drain or sewer. Before you install any new plumbing to an old system, verify the size and ratings of the system. If your new work will rely on the old plumbing to function properly, the old plumbing must meet current code requirements.

Beware of the change-in-use regulations. If you do commercial plumbing, this regulation can have a particularly serious effect on your plumbing costs and methods. If the use of a building changes, the plumbing may also have to change. The change-in-use regulations come into play most often on commercial properties, but they can affect a residential building.

CODE UPDATE

Pipes that pass through concrete or cinder walls and floors must be protected from corrosion.

CODE UPDATE

Pipe, other than cast-iron and galvanized pipe, must be protected from damage when it penetrates sections of concealed areas. This type of protection is required if the pipe is within 1.5 inches of the finished surface of the concealed area. Steel plates with a thickness of 0.0575 inch must be installed to protect pipes when protection is required.

Assume for a moment that you receive a request to install a three-bay sink in a convenience store. You discover that the store's owner is having the sink installed so that he may prepare food for a new deli in the store. This store has never been equipped for food preparation and service. What complications could arise from this situation? First, zoning may not allow the store to have a deli. Second, if the use is allowed, the plumbing requirements for the store may soar. There could be a need for grease traps, indirect wastes, and a number of other possibilities. When you are asked to perform plumbing that involves a change of use, investigate your requirements before committing to the job.

The remaining general regulations of the plumbing code are easily understood. By reading your local code book you should have no trouble in understanding the regulations.

PERMITS

Permits are generally required for many types of plumbing jobs. When a permit is required, it must be obtained before any work is started. Minor repairs and drain cleaning do not require the issuance of permits. In most cases plumbing permits can only be obtained by master plumbers or their

Fast Fact

When a permit is required, it must be obtained before any work is started.

CODE UPDATE

Urinals installed for use by the public or employees must occupy a separate area with walls or partitions to provide privacy. Check your local code requirements for heights and widths of privacy shields.

agents. In some cases, however, homeowners may be granted plumbing permits for work to be done by themselves in their own homes. Permits are obtained from the local code-enforcement office, and that office provides the necessary forms for permit applications.

The information required to obtain a permit will vary from jurisdiction to jurisdiction. You may be required to submit plans, riser diagrams, and specifications for the work to be performed. At a minimum, you will probably be required to adequately describe the scope of work to be performed, the location of the work, the use of the property, and the number and type of fixtures being installed.

The amount of information required to obtain a permit is determined by the local code officer. It is not unusual for the code officer to require two sets of plans and specifications for the work to be performed. The detail of the plans and specs is also left up to the judgment of the code officer. Requirements may include details of pipe sizing, grade, fixture units, and any other information the code officer may deem pertinent.

If your work will involve working with a sewer or water service, expect to be asked for a site plan. The site plan should show the locations of the water service and sewer. If you will be working with a private sewage-disposal system, its location should be indicated on the site plan. Once your plans

Fast Fact

It is possible to obtain a partial permit when time is of the essence. This is a permit that approves a portion of work proposed for completion. Just remember that there is risk involved.

Fast Fact

The code book is a guide, not the last word. The last word comes from the code-enforcement officer. This is an important fact to remember.

are approved, any future changes to the plans must be submitted to and approved by the code officer.

Plumbing permits must be signed by the code officer or an authorized representative. If you submitted plans with your permit application, the plans will be labeled with appropriate wording to establish that they have been reviewed and approved. If it is later found that the approved plans contain a code violation, the plumbing must be installed according to code requirements, regardless of whether the approved plans contain a nonconforming use. Most jurisdictions require a set of approved plans to be kept on the job site and available to the code officer at all reasonable times.

If plans are required, they must be approved before a permit is issued. All fees associated with the permit must be paid prior to the issuance of the permit. These fees are established by local jurisdictions. After a permit is issued, all work must be done in the manner presented during the permit-application process.

It is possible to obtain a partial permit when time is of the essence. This is a permit that approves a portion of work proposed for completion. Just remember that there is risk involved. Assume, for example, that you obtained permission and a permit to install your underground plumbing but had not yet been issued a permit for the remainder of the job. As winter approaches, you decide to in-

CODE UPDATE

Shower pan liners must be tested for leaks before being used to build showers on. The liner must hold a minimum of two inches of potable water for not less than 15 minutes.

Fast Fact

Every job that requires a permit also requires an inspection. Many jobs require more than one inspection.

stall your groundwork so that the concrete floor can be poured over the plumbing, before freezing conditions arrive. This is a good example of how and why partial approvals are good but can result in problems.

You have installed your underground plumbing, and the concrete is poured. After a while you are notified by the code officer that the proposed above-grade plumbing is not in acceptable form and will require major changes. These changes will affect the location and size of your underground plumbing. What do you do now? Well, you are probably going to spend some time with a jackhammer or concrete saw. The underground plumbing must be changed, or the above-ground plumbing must be redesigned to work with the groundwork. In either case, you have trouble and expense that would have been avoided if you had not acted on a partial approval. Partial approvals have their place, but use them cautiously.

How can a plumbing permit become void? If you do not begin work within a specified time, normally six months, your permit will be considered abandoned. When this happens, you must start the entire process over again to obtain a new permit. Plumbing permits can be revoked by the code official. If it is found that facts given during the permit application were false, the permit may be revoked. If work stops for an extended period of time, normally six months, a permit may be suspended.

CODE UPDATE

Condensate disposal from cooling coils and evaporators must be conveyed from drain pans to approved places of disposal. The grading for a condensate drain is to be a minimum of ⅛ inch per foot. Streets, alleys, and other areas where condensate drainage may cause a nuisance are not suitable discharge locations.

CODE OFFICERS

Code enforcement is generally performed on the local level by local officers. These individuals are frequently referred to as inspectors. It is their job to interpret and enforce the regulations of the plumbing code. Since code-enforcement officers have the duty of interpreting the code, there may be times when a decision is reached that appears to contradict the code book. The book is a guide, not the last word. The last word comes from the code-enforcement officer. This is an important fact to remember. Regardless of how you interpret the code, it is the code officer's decision that is final.

Code Approval

Every job that requires a permit also requires an inspection and code approval. Many jobs require more than one inspection. In the plumbing of a new home there may be as many as four inspections. One inspection would be for the sewer and water-service installation. Another inspection might be for underground plumbing. Then you would have a rough-in inspection for the pipes that are to be concealed in walls and ceilings. Then, when the job is done, there will be a final inspection.

These inspections must be done while the plumbing work is visible. A test of the system is generally required, with pressure from either air or water. Normally the inspection is done by the local code officer, but not always. The code officer may accept the findings of an independent inspection service. Any test results submitted to a code office from an independent testing agency must come from an agency that is approved by the code jurisdiction. Before independent inspection results will be accepted, the inspection service must be approved by the code officer. This is also the case when independent inspection services are used to inspect prefabricated construction.

Plumbing inspectors are generally allowed the freedom to inspect plumbing at any time during normal business hours. These inspectors cannot en-

CODE UPDATE

The minimum allowable diameter for a condensate drain is ¾ inch.

Fast Fact

If you feel you have received an unfair ruling from a code officer, you may make a formal request to an appeal board to have the decision changed.

ter a property without permission unless they obtain a search warrant or other proper legal authority. Permission for entry is frequently granted by the permit applicant when the permit is signed.

Backflow preventers are to be inspected annually. Both backflow assemblies and air gaps must be inspected to ensure that they will operate properly. Backflow preventers and double check-valve assemblies must be inspected when they are installed, immediately after any repairs or relocation, and on a routine annual basis. All testing and inspection must be done in accordance with the requirements of local codes. Other devices falling under these same testing requirements include:

- Reduced pressure principle backflow-preventer assemblies
- Double check-valve assemblies
- Pressure vacuum breaker assemblies
- Reduced pressure-detector fire-protection backflow-prevention assemblies
- Double check-detector fire-protection backflow preventers
- Hose-connection backflow preventers
- Spillproof vacuum breakers.

What Inspectors Look For

When plumbing inspectors look at a job, they are looking at many aspects of the plumbing. They will inspect to see that the work is installed in compliance with the code and in a way that it is likely to last for its normal lifetime. Inspectors will check to see that all piping is tested properly and that all plumbing is in good working order.

CODE UPDATE

Auxiliary and secondary drain systems may be required in some condensate drainage designs. An auxiliary pan must have its own drain. The point of discharge must be conspicuous so that occupants or others will notice the discharge and bring it to the attention of proper people. Drainage pans are required to have a minimum depth of 1.5 inches.

Plumbing Police

Plumbing inspectors can be considered the plumbing police. These inspectors have final authority over any plumbing-related issue. If plumbing is found to be in violation of the code, plumbing inspectors may take several forms of action to rectify the situation.

Normally, inspectors will advise the permit holder of the code violations and allow a reasonable time for correction. This advice will come in the form of written documents and will be recorded in an official file. If the violations are not corrected, the code officer will take further steps. Legal counsel may be consulted. After a legal determination is made, action may be taken against the permit holder in violation of the plumbing code. This could involve cash fines, license suspension, license revocation, and in extreme cases jail.

Code officers have the power to issue a stop-work order. This order requires all work to stop until code violations are corrected. These orders are not used casually; they are used when an immediate danger is present or possible. Code officers do have a protocol to follow in the issuance of stop-work orders. If you ever encounter a stop-work order, stop working. These orders are serious, and violation can deliver more trouble than you ever imagined.

When code officers inspect a plumbing system and find it to be satisfactory, they will issue an approval of the system. This allows the pipes to be concealed and the system to be placed into operation. In certain circumstances code officers may issue temporary approvals. These approvals are issued for portions of a plumbing system when conditions warrant them.

CODE UPDATE

Water-level monitoring devices are required on condensate drainage systems when down-flow units and all other coils do not have a secondary drain. The monitor must be equipped to shut down if the primary drain becomes restricted. Devices installed in drain lines are not permitted.

When a severe hazard exists, plumbing inspectors have the power to condemn property and force occupants to vacate. This power would only be used under extreme conditions where a health or safety hazard was present.

Code officers are empowered to authorize the disconnection of utility services to a building in the case of an emergency or when there is a need to eliminate an immediate danger to life or property. Code officers must attempt to notify residents and owners of the intent to disconnect utility service when time allows. If residents and owners cannot be notified prior to disconnection, they shall be notified in writing as soon as practical thereafter.

A Code Officer's Decision?

If you feel you have received an unfair ruling from a code officer, you may make a formal request to an appeal board to have the decision changed. Your reasons for an appeal must be valid and pertinent to specific code requirements. Your appeal could be based on what you feel is an incorrect interpretation of the code. If you feel the code does not apply to your case, you have reason for an appeal. There are other reasons for appeal, but you must specify why the appeal is necessary and how the decision you are appealing is incorrect.

CODE UPDATE

Toilet facilities are required in all occupancies and can be either separate or combined with public facilities.

HEALTH AND SAFETY

Health and safety are two key issues in the plumbing code. These two issues are, by and large, the reasons for the plumbing code's existence. The plumbing code is designed to assure health and safety to the public. Public health can be endangered by faulty or improperly installed plumbing. Code officers have the power and duty to condemn a property where severe health or safety risks exist. It is up to the owner of each property to maintain the plumbing in a safe and sanitary manner.

When it comes to safety, there are many more considerations than just plumbing pipes. Most safety concerns arise in conjunction with plumbing but not from the plumbing itself. It is far more likely that a safety hazard will result from the activities of a plumber on other aspects of a building. An example would be cutting so much of a bearing timber that the structure becomes unsafe. Perhaps a plumber removes a wire from an electric water heater and leaves it exposed and unattended; this could result in a fatal shock to someone. The list of potential safety risks could go on for pages, but you get the idea. It is your responsibility to maintain safe and sanitary conditions at all times.

A part of maintaining sanitary conditions includes the use of temporary toilet facilities on job sites. It is not unusual for the plumbing code to require toilet facilities to be available to workers during the construction of buildings. These facilities can be temporary, but they must be sanitary and available.

Trusses must not be cut, drilled, notched, spliced, or otherwise altered in any way without the written concurrence and approval of a registered design professional. Additionally, trusses must not be used to support addi-

CODE UPDATE

Tables are often used to determine the number and type of plumbing fixtures that are required in a building. In the event that your calculations for the number of people to use the facility is a fractional number, round up to the next highest number for your sizing procedures.

Trade Tip

Trusses must not be cut, drilled, notched, spliced, or otherwise altered in any way without the written concurrence and approval of a registered design professional.

tional loads, such as mechanical systems and equipment, without verification that the truss is capable of supporting the additional load safely.

Pipe Protection

It is the plumber's responsibility to protect plumbing pipes. This protection can take many forms. Here we are going to look at the basics of pipe protection. You will gain insight into pipe-protection needs that you may have never considered before.

Backfilling

When backfilling over a pipe, you must take measures to prevent damage to the pipe. The damage can come in two forms, immediate damage and long-term damage. If you are backfilling with material that contains large rocks or other foreign objects, the weight or shape of the rocks and objects may puncture or break the pipe. The long-term effect of large rocks next to a pipe could result in stress breaks. It is important to use only clean backfill material when backfilling pipe trenches.

Even the weight of a large load of backfill material could damage the pipe or its joints. Backfill material should be added gradually. Layers of backfill should not be more than 6 inches deep before they are compacted. Each layer

Trade Tip

It is important to use only clean backfill material when backfilling pipe trenches.

CODE UPDATE

Directional signage is required to be located in a corridor or aisle, at the entrance to the facilities.

of this backfill should be compacted before the next load is dumped. Backfill material both under and beside a pipe must be compacted for pipe support.

The crown of a buried pipe must be covered by at least 12 inches of tamped earth. Backfill material is to be installed under and beside a pipe to provide compacted support. All materials must be installed in accordance with the most restrictive recommendations, whether they are dictated by the code or the pipe manufacturer. However, if there is a conflict between the code and the manufacturer's installation instructions, the provisions of the code apply. The exception to this rule is when the manufacturer's instructions are more restrictive than those of the code. In other words, you must abide by the instructions or code language that is the most restrictive.

Flood Protection

If a plumbing installation is made in an area subject to flooding, special precautions must be taken. High water levels can float pipes and erode the earth around them. If your installation is in a flood area, consult your local code officer for the proper procedures to protect your pipes.

Floodproofing must be in accordance with the requirements of the International Building Code.

Essentially, plumbing systems installed in structures that are built in flood-hazard areas must be located above the design flood elevation. There are, however, exceptions to this rule. Some systems can be installed below the design flood elevations under certain circumstances. Plumbing equipment, supplies, and devices located below a design flood elevation must be designed and installed in a way that prevents water from entering or accumulating within the components and systems. A system must be constructed to resist hydrostatic and hydrodynamic loads and stresses, including the effects of buoyancy, during any occurrence of flooding. The types of systems and equipment that may be allowed to be installed below a design flood elevation include:

- Water-service pipes
- Pump seals in individual water-supply systems where the pump is located below the design flood elevation
- Covers on potable water wells must be sealed, except where the top of the casing well or pipe sleeve is at least 1 foot above the design flood elevation
- Sanitary-drainage piping
- Storm-drainage piping
- Vent systems
- Manhole covers must be sealed, except where elevated to or above the design flood elevation
- Water heaters
- All other plumbing fixtures, faucets, fixture fittings, and piping systems and equipment

Plumbing systems, pipes, and fixtures located in a flood-hazard area that is subject to high-velocity wave action must not be mounted on or penetrate through walls intended to break away under flood loads.

Penetrations

When a pipe penetrates an exterior wall, it must pass through a sleeve. The sleeve should be at least two pipe sizes larger than the pipe passing through it. Once the pipe is installed, the open space between the pipe and the sleeve should be sealed with a flexible sealant. By caulking around the pipe you

CODE UPDATE

Floor drains must have removable strainers and the drain must be of such a type to allow proper cleaning as needed. The drain must be accessible.

Trade Tip

Pipe connections can require a variety of adapters when combining pipes of different types. It is important to use the proper methods when making any connections, especially when you are mating different types of pipes.

eliminate the invasion of water and rodents. If the penetration is through a fire-resistive material, the space around the sleeve must be sealed with an approved, fire-resistive material.

Freezing

Pipes must be protected against freezing conditions. Outside, this means placing the pipe deep enough in the ground to avoid freezing. The depth will vary with geographic locations, but your local code officer can provide you with minimum depths. Normally, exterior water-supply piping must be installed at least 6 inches below the frost line and not less than 12 inches below finished grade level. Aboveground pipes in unheated areas must be protected with insulation or other means of protection from freezing.

Corrosion

Pipes that tend to be affected by corrosion must be protected. This protection can take the form of a sleeve or a special coating applied to the pipe. For example, copper pipe can have a bad reaction when placed in contact with concrete and should be protected with a sleeve. The sleeve can be a foam insulation or some other type of noncorrosive material. Some soils are capable of corroding pipes. If corrosive soil is suspected, you may have to protect entire sections of underground piping.

Seismic Zones

Pipes hung in seismic zones 3 and 4 that are hubless cast iron 5 inches in diameter and larger and are suspended in exposed locations over public or high traffic areas must be supported on both sides of the coupling if the length of the pipe exceeds 4 feet.

Firestop Protection

Any DWV or storm-water piping penetrations of fire-resistant materials and enclosures must be protected in accordance with all code requirements. Plans and specifications must detail clearly how penetrations of fire-resistive materials and spaces will be firestopped for adequate protection before a permit will be issued. All firestopping materials must be code-approved. There are a number of ratings that pertain to firestopping. For example, an F rating is the time period that the penetration firestop system limits the spread of fire through the penetration when tested in accordance with ASTM E 814. If you run across a T rating, you are dealing with the time period that the penetration firestop system, including the penetrating item, limits the maximum temperature rise of 325°F above its initial temperature through the penetration on the nonfire side when tested in accordance with ASTM E 814.

Combustible Installations

All ABS and PVC DWV piping installations must be protected in accordance with the appropriate fire-resistant rating requirements in the building code. These list the acceptable area, height, and type of construction for use in specific occupancies to assure compliance and integrity of the fire-resistance rating prescribed.

All penetrations must be protected by an approved penetration firestop installation. The systems must have an F rating of at least 1 hour, but not less than the required fire-resistance rating of the assembly being penetrated. Systems that protect floor penetrations must have a T rating of at least one hour but not less than the required fire-resistance rating of the floor being penetrated. Floor penetrations contained within the cavity of a wall at the location of the floor penetration do not require a T rating. No T rating is

CODE UPDATE

Commercial food waste grinders require a drain with a minimum diameter of 1.5 inches and must be connected and trapped separately from other fixtures or sink components.

required for floor penetrations by piping that is not in direct contact with combustible material.

When piping penetrates a rated assembly, noncombustible piping must not be connected to combustible piping unless it can be shown that the transition complies with all code requirements. Before any piping is concealed, the installation must be inspected and approved.

Noncombustible Installations

The basic rules that we just covered for plastic pipe apply to metallic pipe. There are, however, some differences. For example, concrete, mortar, or grout may be used to fill the annular spaces around cast iron, copper, or steel piping. The nominal diameter of the penetrating item should not exceed 6 inches, and the opening size should not exceed 144 square inches. Thickness of the firestop should be the same as the assembly being penetrated. Unshielded couplings are not to be used to connect noncombustible piping unless it can be demonstrated that the fire-resistive integrity of the penetration is maintained.

Inspections

The inspection process for firestopping is handled by the administrative authority. An external examination covers the assembly type, insulation type and thickness, type and size of any sleeve, type and size of penetrant, size of opening, orientation of penetrant, annular space, and rating. The approved drawing should be submitted for the compliance inspection.

An internal examination is usually made with a contractor present and prepared to make repairs. The contractor is asked to cut into the firestop enough to reveal the type and backing materials and the type and amount of the material. Then the contractor repairs the cut and the code officer inspects the repair. Assuming that all goes well, the inspection is then approved.

Connections

Pipe connections can require a variety of adapters when combining pipes of different types. It is important to use the proper methods when making any connections, especially when you are mating different types of pipes. There are many universal connectors available to plumbers today. These

CODE UPDATE

Piping for shower-head risers must be secured in an approved manner.

special couplings are allowed to connect a wide range of materials. Threaded joints on iron pipe and fittings must have standard-taper pipe threads. Threads on tubing must be of an approved type. When flared joints are made for soft copper tubing, they must be made with approved fittings. The tubing must be reamed to the full inside diameter, resized to round, and expanded with a proper flaring tool.

Male and female adapters have long been an acceptable method of joining opposing materials, but today the options are much greater. You can use compression fittings and rubber couplings to match many types of materials to each other. Special insert adapters allow the use of plastic pipe with bell-and-spigot cast iron.

Condensate Disposal

Liquid combustion byproducts of condensing appliances must be collected and discharged to an approved plumbing fixture or disposal area in accordance with the manufacturer's installation instructions. All piping used to handle the condensate must be made of an approved, corrosion-resistant material and must not be smaller than the drain connection on the appliance. Condensate piping must maintain a minimum horizontal slope in the direction of discharge that is at least one-eighth unit vertical in 12 units horizontal.

Equipment containing evaporators of cooling coils must be provided with a condensate drainage system. The system must be designed and installed in accordance with code requirements. Condensate drainage from all cooling coils and evaporators must be conveyed from the drain-pan outlet to an approved place of disposal. It must not be allowed to dump into a street, alley, or other area so as to cause a nuisance.

The components used to create a condensate drainage system may be made of any of the following materials:

- Galvanized steel
- Copper
- Polybutylene
- Cross-linked polyethylene
- Polyethylene
- ABS
- PVC
- CPVC
- Cast iron.

All components used to build a condensate drainage system must be rated for the pressure and temperature requirements of the system. Waste lines must not be less than 0.75 inch in internal diameter. The size is not allowed to decrease from the drainage-pan connection to the place of disposal. If multiple drainpipes are connected to a single disposal pipe, the piping must be sized to meet total flow requirements. Horizontal sections of piping must be installed uniformly and with proper pitch.

If there is risk of damage to any building component, a secondary drainage system is required. There are different ways to comply with this requirement. One way is to install an auxiliary pan with a separate drain under the coils on which condensation will occur. This pan must discharge to a conspicuous location to alert people that the primary drain is not functioning. The pan must have a minimum depth of 1.5 inches and it must be not less than 3 inches larger than the unit or the coil dimensions in width and length. The pan must be made of corrosion-resistant material. When metal pans are used, they must have a minimum thickness of 0.0276 inches.

Another option is to use a separate overflow drainpipe that is connected to the main drain pan. This drain, too, must discharge into a conspicuous point. The secondary drain is required to connect to the drain pan at a location higher than the primary discharge pipe.

A third option is to use an auxiliary drain pan without a separate drain. This pan must be equipped with a water-level detection device. The device will be designed and installed to cut off the equipment being served if the water reaches a high level.

CODE UPDATE

Shower liners must not be perforated within 1 inch above the finished threshold. The liners are required to be graded at a rate of ¼ inch to the foot toward the drain.

All condensate drains must be trapped in accordance with the appliance manufacturer's recommendations. Equipment efficiencies must conform to the International Energy Conservation Code.

Testing

Drainage systems are to be tested with water. This test can be done either on sections of piping or on an entire system. When a system is tested, all openings must be tightly closed, with the exception of the highest opening. The system must be filled to a point where it overflows. If a section of piping is being tested, such as the underground piping, the highest opening must be at least 10 feet high. Eventually all piping must be tested. A test is to last for a minimum of 15 minutes, with the system retaining all water contained during the test.

A final test on a DWV system may involve the use of smoke. This requirement is at the option of the code officer. If a smoke test is required, all traps are to be filled with water and then a pungent, thick smoke is introduced into the plumbing system. Once the smoke is evident at roof vent terminals, the vent openings must be capped. A pressure equal to a 1-inch water column must be maintained for a period of at least 15 minutes.

Tests that require a pressure of 10 psi or less must be done with a test gauge that has increments of 0.10 psi or less. If a test requires a pressure of greater than 10 but less than 100 psi, it must be done with a test gauge calibrated in increments of 1 psi or less. When a pressure of more than 100 psi is required, the test gauge must have increments of 2 psi or less.

WORKING WITH THE SYSTEM INSTEAD OF AGAINST IT

Code officers are expected to enforce the regulations set forth by the code. Plumbers are expected to work within the parameters of the code. Naturally, plumbers and code officers will come into contact with each other on a regular basis. This contact can lead to some disruptive actions.

The plumbing code is in place to help people, not hurt them. It is not meant to ruin your business or to place you under undue hardship in earning a living. It really is no different from our traffic laws. The traffic laws are there to protect all of us, but some people resent them. Some plumbers resent the plumbing code. They view it as a vehicle for the local jurisdiction to make more money while they, the plumbers, are forced into positions to possibly make less money.

When you learn to understand the plumbing code and its purpose, you will learn to respect it. You should respect it; it shows the importance of your position as a plumber to the health of our entire nation. Whether you agree with the code or not, you must work within its constraints. This means working with the inspectors.

When inspectors choose to play hardball, they hold most of the cards. If you develop an attitude problem, you may be paying for it for years to come. Even if you know you are right on an issue, give the inspector a place to escape; nobody enjoys being ridiculed in his or her profession.

The plumbing code is largely a matter of interpretation. If you have questions, ask your code officer for help. Code officers are generally more than willing to give advice. It is only when you walk into their offices with a chip on your shoulder that you are likely to hit the bureaucratic wall. Like it or not, you must learn to comply with the plumbing code and to work with code officers. The sooner you learn to work with them on amicable terms, the better off you will be.

Little things can mean a lot. Apply for your permit early. This prevents the need to hound the inspector to approve your plans and issue your permit.

CODE UPDATE

The maximum temperature of hot water in bathtubs and whirlpool bathtubs is 120°F.

Many jurisdictions require at least 24 hours advance notice for an inspection request, but even if your jurisdiction doesn't have this rule, be considerate and plan your inspections in advance. By making life easier for the inspector, you will be helping yourself.

CHAPTER

4

Fixtures, Faucets, and Fixture Fittings

There is much more to fixtures than meets the eye. Fixtures are a part of the final phase of plumbing. When you are planning a plumbing system, you must know which fixtures and types are required and how they must be installed. This chapter will guide you through the many fixtures available and how they may be used.

FIXTURES

The number and type of fixtures required depends on local regulations and the use of the building where they are being installed. Your code book will provide you with information on which types and how many of each are required. These requirements are based on the use of the building housing the fixtures and the number of people that may be using the building. Let me give you some examples.

Single-Family Home

When you are planning fixtures for a single-family residence, you must include certain fixtures. If you choose to install more than the minimum, that's fine, but you must install the minimum number of required fixtures. The minimum number and type of fixtures for a single-family dwelling are as follows:

- One toilet
- One lavatory
- One bathing unit
- One kitchen sink
- One washing-machine hookup.

Multi-Family Buildings

The minimum requirements for a multi-family building are the same as those for a single-family dwelling, but each dwelling in the building must be equipped with the minimum fixtures. There is one exception—the laundry hookup. With a multi-family building, laundry hookups are not required in each dwelling unit. A laundry hookup for common use may be required when there are 20 or more dwelling units. For each interval of 20 units, you must install a laundry hookup when this code is in effect.

For example, in a building with 40 apartments, you would have to provide two laundry hookups. If the building had 60 units, you would need three hookups. This ratio is not always the same. Sometimes a dwelling-unit interval is 10 rental units, and local requirements could require one hookup for every 12 rental units but no fewer than two hookups for buildings with at least 15 units.

Public Assembly

for businesses and places of public assembly such as nightclubs, the ratings are based on the number of people likely to use the facilities. In a nightclub, the minimum requirements are usually as follows:

- One toilet for every 40 people
- Fixtures located in a unisex toilet or bathing room can be counted in determining the minimum required number of fixtures for assembly and mercantile occupancies only

Fast Fact

When installing separate bathroom facilities, the number of required fixtures will be divided equally between the two sexes, unless there is cause for and approval of a different appropriation. Separate facilities are not required for private facilities.

- One lavatory for every 75 people
- One service sink
- One drinking fountain for every 500 people
- Zero bathing units.

Drinking fountains are not required in establishments such as restaurants, where water is served. When drinking fountains are required, bottled-water dispensers can be substituted for up to 50 percent of the requirement.

Day-Care Facilities

The minimum number of fixtures for a day-care facility are usually:

- One toilet for every 15 people
- One lavatory for every 15 people
- One bathing unit for every 15 people
- One service sink
- One drinking fountain every 100 people.

Some local codes only require the installation of toilets and lavatories in day-care facilities. This information will be found in your local code and will cover all the normal types of building uses.

In many cases, facilities will have to be provided in separate bathrooms to accommodate each sex. When installing separate bathroom facilities, the number of required fixtures is divided equally between the two sexes, unless there is cause for and approval of a different appropriation. Separate facilities are not required for private facilities.

Fast Fact

There are some special regulations pertaining to employee and customer facilities.

Some types of buildings do not require separate facilities. For example, some jurisdictions do not require residential properties or small businesses where less than 15 employees work or where less than 15 people are allowed in the building at the same time to have separate facilities.

Local codes may also not require separate facilities in the following buildings: offices with less than 1200 square feet, retail stores with less than 1500 square feet, restaurants with less than 500 square feet, self-serve laundries with less than 1400 square feet, and hair salons with less than 900 square feet. Separate facilities are not required for private facilities. Mercantile occupancies in which the maximum occupant load is 50 or less and food and beverages are not being served are not required to have separate facilities.

SPECIAL REGULATIONS

There are some special regulations pertaining to employee and customer facilities. For employees, toilet facilities must be available to employees within a reasonable distance and with relative ease of access. The general code requires these facilities to be in the immediate work area; the distance an employee is required to walk to the facilities may not exceed 500 feet. The facilities must be located in a manner so that employees do not have to negotiate more than one set of stairs for access to the facilities. There are some exceptions to these regulations, but in general these are the rules.

It is expected that customers of restaurants, stores, and places of public assembly will have toilet facilities. This is usually based on buildings capable of holding 150 or more people. Buildings with an occupancy rating of less than 150 people are not normally required to provide toilet facilities unless the building serves food or beverages. When facilities are required, they may be placed in individual buildings or, in a shopping-mall situation, in a common area not more than 500 feet from any store or tenant space. These central toilets must be placed so that customers will not have to use more than one set of stairs to reach them.

Toilet facilities in buildings other than assembly or mercantile cannot be installed more than one story above or below the employees' regular working area, and the path of travel to the toilet facilities cannot exceed 500 feet. There is a potential exception for maximum travel distance when the building is used for factory or industrial purposes.

When toilet facilities for employees are located in covered malls, the travel distance must not exceed 300 feet. There are exceptions to this rule, so check your local requirements. Facilities in covered malls are based on total square footage. Toilet facilities must be installed in each individual store or in a central toilet area located no more than 300 feet from the source of travel for individuals using the facility. Travel distance is measured from the main entrance of any store or tenant space.

Some jurisdictions use a square-footage method to determine minimum requirements in public places. For example, retail stores are rated as having an occupancy load of one person for every 200 square feet of floor space. This type of facility is required to have separate facilities when the store's square footage exceeds 1500 square feet. A minimum of one toilet is required for each facility when the occupancy load is up to 35 people. One lavatory is required in each facility for up to 15 people. A drinking fountain is required for occupancy loads up to 100 people. Drinking fountains may not be installed in public restrooms.

Pay-bathroom facilities are allowed, but these facilities must be installed in addition to the minimum plumbing requirements for free facilities.

Trade Tip

Accessible fixtures are not cheap; you cannot afford to overlook them when bidding a job. The plumbing code normally requires specific minimums for accessible fixtures in certain circumstances.

Public plumbing facilities must be identified with legible signs for each sex. The signage must be readily visible and near the entrance to the toilet facility.

Accessible Fixtures

Accessible fixtures are not cheap; you cannot afford to overlook them when bidding a job. The plumbing code normally requires specific minimums for accessible fixtures in certain circumstances. It is your responsibility to know when such facilities are required. There are also special regulations pertaining to how they shall be installed.

When you are dealing with accessible plumbing, you must combine the local plumbing code with the local building code. These two codes work together in establishing the minimum requirements for accessible plumbing facilities. When you step into the field of accessible plumbing, you must play by a different set of rules. Accessibility is like a different code all unto itself.

Accessible-Fixture Requirements

Most buildings frequented by the public are required to have accessible plumbing fixtures. The following examples are based on general code requirements.

Single-family homes and most residential multi-family dwellings are exempt from accessibility requirements. A rule-of-thumb for most public buildings is the inclusion of one accessible toilet and one lavatory.

Hotels, motels, inns, and the like are required to provide an accessible toilet, lavatory, bathing unit, and kitchen sink, where applicable. Drinking fountains may also be required. Drinking fountains must not be installed in public restrooms. This provision will depend on the local plumbing and building codes. If plumbing a gang-shower arrangement, such as in a

Trade Tip

When you are dealing with accessible plumbing, you must combine the local plumbing code with the local building code.

school gym, at least one of the shower units must be accessible. Door sizes and other building-code requirements must be observed when dealing with accessible facilities. There are local exceptions to these rules; check with your local code officers for current local regulations.

Accessible Plumbing

When installing accessible plumbing facilities, you must pay attention to the plumbing code and the building code. In most cases, approved blueprints will indicate the requirements of your job, but in rural areas you may not enjoy the benefit of highly detailed plans and specifications. At the final inspection, the plumbing must pass muster along with the open space around the fixtures. If the inspection is failed, your pay is held up and you are likely to incur unexpected costs. This section will apprise you of what you need to know.

Accessible Toilet Stalls

When you think of installing an accessible toilet, you probably think of a toilet that sits high off the floor. But do you think of the grab bars and partition dimensions required around the toilet? Some plumbers don't, but they should. The door to a privacy stall for an accessible toilet must provide a minimum of 32 inches of clear space for wheelchair access.

The distance between the front of the toilet and the closed door must be at least 48 inches. It is mandatory that the door open outward, away from the toilet. Think about it: how could a person in a wheelchair close the door if the door opened in to the toilet? These facts may not seem like your problem, but if your inspection doesn't pass, you don't get paid.

The width of a water-closet compartment for accessible toilets must be a minimum of 5 feet. The length of the privacy stall must be at least 56 inches for wall-mounted toilets and 59 inches for floor-mounted models. Unlike regular toilets, which require a rough-in of 15 inches to the center of the drain from a sidewall, accessible toilets require the rough-in to be at least 18 inches off the sidewall.

Next are the required grab bars. Sure, you may know that grab bars are required, but do you know the mounting requirements for the bars? Two bars are required for each handicap toilet. One bar should be mounted on the back wall and the other on the side wall. The bar mounted on the back wall must be at least 3 feet long. The first mounting bracket of the bar must be mounted no more than 6 inches from the side wall. The bar must extend at least 24 inches past the center of the toilet's drain.

The bar mounted on the sidewall must be at least 42 inches long. The bar should be mounted level and with the first mounting bracket located no more than 1 foot from the back wall. The bar must be mounted on the side wall that is closest to the toilet. This bar must extend to a point at least 54 inches from the back wall. If you do your math, you will see that a 42-inch bar is pushing the limits on both ends. A longer bar will allow more assurance of meeting the minimum requirements.

When a lavatory will be installed in the same toilet compartment, the lavatory must be installed on the back wall. The lavatory must be installed in such a way that its closest point to the toilet is no less than 18 inches from the center of the toilet's drain. When a privacy stall of this size and design is not suitable, another way to size the compartment to house an accessible toilet and lavatory is available. There may be times when space restraints will not allow a stall with a width of 5 feet. In these cases, you may position the fixture differently and use a stall with a width of only 3 feet. In these situations, the width of the privacy stall may not exceed 4 feet.

The depth of the compartment must be at least 66 inches when wall-mounted toilets are used. The depth extends to a minimum of 69 inches with the use of a floor-mounted water closet. The toilet requires a minimum distance from side walls of 18 inches to the center of the toilet drain. If the compartment is more than 3 feet wide, grab bars are required, with the same installation as described before.

If the stall is made at the minimum width of 3 feet, grab bars, with a minimum length of 42 inches, are required on each side of the toilet. These bars must be mounted no more than 1 foot from the back wall, and they must extend a minimum of 54 inches from the back wall. If a privacy stall is not used, the side-wall clearances and the grab-bar requirements are the same as listed in these two examples. To determine which set of rules to use, you must assess the shape of the room when no stall is present.

If the room is laid out as in the first example, use the guidelines for grab bars listed there. If, on the other hand, the room tends to meet the description of the last example, use the specifications in that example. In both cases, the door to the room may not swing into toilet area.

People with Less Physical Ability

Accessible fixtures are specially designed for people with less physical ability than the general public. The differences may appear subtle, but they are important. Let's look at the requirements a fixture must meet to be considered an accessible fixture.

In assembly and mercantile occupancies, unisex toilet and bathing rooms must be provided in accordance with the local code. An accessible unisex toilet room is required when an aggregate of six or more male or female water closets are required. In buildings with mixed-use occupancy, only the water closets required for the assembly or mercantile occupancy are used to determine the unisex-toilet room requirement. A unisex bathing room is required in recreational facilities where separate-sex bathing rooms are provided. There is an exception: When a separate-sex bathing room has only one shower or bathtub, a unisex bathing room is not required.

Accessible toilet facilities must comply with all code requirements. The facilities must consist of only one water closet and only one lavatory. Unisex bathing rooms are to be considered a unisex toilet room. As usual, there are exceptions: A separate-sex toilet room that contains no more than two water closets, no urinals, or one water closet and one urinal is considered a unisex toilet room.

When unisex toilet and bathing rooms are installed, they must be located on an accessible route. The rooms must not be located more than one story above or below the separate-sex toilet room. Travel distance is not to exceed 500 feet.

Unisex toilet rooms installed in passenger transportation facilities and airports must have a travel route from separate-sex toilet rooms that does not require passage through security checkpoints.

A clear floor space of not less than 30 inches by 48 inches is required beyond the area of the door swing when a door opens into a unisex toilet or bathing room. Doors providing privacy for unisex toilet and bathing rooms must be securable from within the room. A sign that complies with code requirements is required to designate a unisex toilet or bathing room. Directional signage must be provided at all separate-sex toilet and bathing rooms indicating the location of the nearest unisex toilet or bathing room.

Accessible Toilets

Accessible toilets have a normal appearance, but they sit higher above the floor than a standard toilet. The toilet will rise to a height of between 16 and 20 inches off the finished floor; 18 inches is a common height for most models. Toilets are required to have a minimum of 30 inches center-to-center between other fixtures and walls. An open space of 21 inches is required in front of the toilet. The same is true for bidets. Urinals require a center-to-center open space of 30 inches. There are many choices in toilet style, including the following:

- Siphon jet
- Siphon vortex
- Siphon wash
- Reverse trap
- Blowout.

Accessible Sinks

Accessible sinks, lavatories, and faucets may appear to be standard fixtures, but their method of installation is regulated and the faucets are often unlike a standard faucet. Sinks and lavatories must be positioned to allow a person in a wheelchair to use them easily.

The clearance requirements for a lavatory are numerous. There must be at least 21 inches of clearance in front of the lavatory. This clearance must extend 30 inches from the front edge of the lavatory or countertop, whichever protrudes the furthest, and to the sides. If you can sit a square box, with a 30-x-30-inch dimension in front of the lavatory or countertop, you have adequate clearance for the first requirement. This applies to kitchen sinks and lavatories.

The next requirement calls for the top of the lavatory to be no more than 35 inches from the finished floor. For a kitchen sink the maximum height is 34 inches. There is also knee clearance to consider. The minimum allowable knee clearance requires 29 inches in height and 8 inches in depth. This is measured from the face of the fixture, lavatory, or kitchen sink. Toe clearance is another issue. A space 9 inches high and 9 inches deep is required as a minimum for toe space. The last requirement deals with hot-water pipes. Any exposed hot-water pipes must be insulated or shielded to prevent users of the fixture from being burned.

Lavatories in employee and public toilet rooms must be located in the same room as the required water closet.

Lavatory Faucets

Accessible faucets frequently have blade handles. The faucets must be located no more than 25 inches from the front edge of the lavatory or counter, whichever is closest to the user. The faucets can use wing handles, single handles, or push buttons to operate, but the operational force required by the user shall not be more than 5 pounds.

Bathing Units

Accessible bathtubs and showers must meet the requirements of approved fixtures, like any other fixture, but they are also required to have special features and installation methods. The special features are required under the code for approved accessible fixtures. The clear space in front of a bathing unit is required to be a minimum of 1440 square inches. This is achieved by leaving an open space of 30 inches in front of the unit and 48 inches to the sides. If the bathing unit is not accessible from the side, the minimum clearance is increased to an area with a dimension of 48 by 48 inches.

Accessible bathtubs are required to be installed with seats and grab bars. The grab bar must have a diameter of at least 1.25 inches. The diameter may not exceed 1.5 inches. All grab bars are installed 1.5 inches from walls. The design and strength of these bars are set forth in the building codes.

The seat may be an integral part of the bathtub, or it may be a removable, after-market seat. The grab bars must be at least 2 feet long. Two of these grab bars are to be mounted on the back wall, one on top of the other. The bars are to run horizontally. The lowest grab bar must be mounted 9 inches above the flood-level rim of the tub. The top grab bar must be mounted a minimum of 33 inches but no more than 36 inches above the finished floor. The grab bars should be mounted near the seat of the bathing unit.

Additional grab bars are required at each end of the tub. These bars should be mounted horizontally and at the same height as the highest grab bar on the back wall. The bar over the faucet must be at least 2 feet long. The bar on the other end of the tub may be as short as 1 foot.

The faucets in these bathing units must be located below the grab bars. They must be able to be operated with a maximum force of 5 pounds. A personal, hand-held shower is required in all accessible bathtubs. The hose for the hand-held shower must be at least 5 feet long.

Two types of showers are normally used for accessibility purposes. The first type allows the user to leave a wheelchair and shower while sitting on a seat. The other style of shower stall is meant for the user to roll a wheelchair into the stall and shower while seated in the wheelchair.

If the shower is intended to be used with a seat, its dimensions should form a square with 3 feet of clearance. The seat should be no more than 16 inches wide and mounted along the side wall. This seat should run the full length of the shower. The height of the seat should be between 17 and 19 inches above the finished floor. There should be two grab bars installed in the shower. These bars should be located between 33 and 36 inches above the finished floor. The bars are mounted in an "L" shape. One bar should be 36 inches long and run the length of the seat, mounted horizontally. The other bar should be installed on the side wall of the shower. This bar should be at least 18 inches long.

The faucet for this type of shower must be mounted on the wall across from the seat. The faucet must be at least 38 inches but not more than 48 inches above the finished floor. A hand-held shower must be installed. The hand-held shower can be in addition to a fixed showerhead, but there must be a hand-held shower on a hose of at least 5 feet in length. The faucet must be able to operate with a maximum force of 5 pounds.

Water Coolers

The distribution of water from a water cooler or drinking fountain must occur at a maximum height of 36 inches above the finished floor. The outlet for drinking water must be located at the front of the unit, and the water must flow upwards for a minimum distance of 4 inches. Levers or buttons to control the operation of the drinking unit may be mounted on the front of the unit or on the side near the front.

Clearance requirements call for an open space of 30 inches in front of the unit and 48 inches to the sides. Knee and toe clearance is the same as required for sinks and lavatories. If the unit is made so that the drinking spout extends beyond the basic body of the unit, the width clearance may be reduced from 48 inches to 30 inches, so long as knee and toe requirements are met.

Standard Fixtures

Standard fixtures must also be installed according to local code regulations. There are space limitations, clearance requirements, and predetermined, approved methods for installing standard plumbing fixtures. First, let's look at the space and clearance requirements for common fixtures.

Placement

Toilets and bidets require a minimum distance of 15 inches from the center of the fixture's drain to the nearest side wall. These fixtures must have at least 15 inches of clear space between the center of their drains and any obstruction, such as a wall, cabinet, or other fixture. With this rule in mind, a toilet or bidet must be centered in a space of at least 30 inches. Most local codes further require that there be a minimum of 21 inches of clear space in front of these fixtures and that when toilets are placed in privacy stalls, the stalls must be at least 30 inches wide and 60 inches deep.

Most codes require urinals to be installed with a minimum clear distance of 15 inches from the center of their drains to the nearest obstacle on either side. When urinals are installed side-by-side, the distance between the centers of their drains must be at least 30 inches. When urinals are substituted for water closets, they cannot comprise more than 67 percent of the requirement. Urinals must also have a minimum clearance of 21 inches in front of them.

Lavatories must be installed so that there is at least 15 inches of clearance from the center of the fixture drain and any side wall or adjacent fixture. A minimum space of 21 inches is required in front of lavatories.

These fixtures, as with all fixtures, must be installed level and with good workmanship. The fixture should normally be set an equal distance from walls to avoid a crooked or cocked installation. All fixtures should be designed and installed with proper cleaning in mind.

Bathtubs, showers, vanities, and lavatories should be placed in a manner to avoid violating the clearance requirements for toilets, urinals, and bidets.

Sealing Fixtures

Some fixtures hang on walls, and others sit on floors. When securing fixtures to walls and floors, there are some rules you must follow. Floor-mounted fixtures, like most residential toilets, should be secured to the floor with the use of a closet flange. The flange is first screwed or bolted to the floor. A wax seal is then placed on the flange, and closet bolts are placed in slots on both sides of the flange. Then the toilet is set into place.

The closet bolts should be made of brass or some other material that will resist corrosive action. The closet bolts are tightened until the toilet will not move from side to side or front to back. In some cases, a flange is not used. When a flange is not used, the toilet should be secured with corrosion-resistant lag bolts.

When toilets or other fixtures are being mounted on a wall, the procedure is a little different. The fixture must be installed on and supported by an approved hanger. The hangers are normally packed with the fixture. The hanger must assume the weight placed in and on the fixture to avoid stress on the fixture itself.

In the case of a wall-hung toilet, the hanger usually has a pattern of bolts extending from the hanger to a point outside the wall. The hanger is concealed in the wall cavity. A watertight joint is made at the point of connection, usually with a gasket ring, and the wall-hung toilet is bolted to the hanger.

With lavatories, the hanger is usually mounted on the outside surface of the finished wall. A piece of wood blocking is typically installed in the wall cavity to allow a solid surface for mounting the bracket. The bracket is normally secured to the blocking with lag bolts. The hanger is put in place, and lag bolts are screwed through the bracket and finished wall into the wood blocking. Then the lavatory is hung on the bracket.

The space where the lavatory meets the finished wall must be sealed. This is true of all fixtures coming into contact with walls, floors, or cabinets. The crevice caused by the fixture mount in the finished surface must be sealed to protect against water damage. A caulking compound, such as silicone, is normally used for this purpose. This seal does more than prevent water damage. It eliminates hard-to-clean areas and makes the plumbing easier to keep free of dirt and germs.

When bathtubs are installed, they must be installed level, and they must be properly supported. The support for most one-piece units is the floor. These units are made to be set into place, leveled, and secured. Other types of tubs, such as cast-iron tubs, require more support than the floor will give. They need a ledger or support blocks placed under the rim, where the edge of the tub meets the back wall.

The ledger can be a piece of wood, such as a wall stud. The ledger should be about the same length as the tub. This ledger is installed horizontally and level. The ledger should be at a height that will support the tub in a level fashion or with a slight incline so excess water on the rim of the tub will run back into the tub. The ledger is nailed to wall studs.

If blocks are used, they are cut to a height that will put the bathtub into the proper position. Then the blocks are placed at the two ends and often in the middle, where the tub will sit. The blocks should be installed vertically and nailed to the stud wall.

When the tub is set into place, the rim at the back wall rests on the blocks or ledger for additional support. This type of tub has feet on the bottom so that the floor supports most of the weight. The edges where the tub meets the walls must be caulked. If shower doors are installed on a bathtub or shower, they must meet the safety requirements set forth in the building codes.

Fast Fact

Incoming water lines must be protected against freezing and backsiphonage.

Showers today are usually one-piece units. These units are meant to be set in place, leveled, and secured to the wall. The securing process for one-piece showers and bathtubs is normally accomplished by placing nails or screws through a nailing flange, which is molded as part of the unit, into the stud walls. If only a shower base is being installed, it must also be level and secure. Shower doors must open enough to allow a minimum of 22 inches of unobstructed opening for egress. Now, let's look at some of the many other regulations involved in installing plumbing fixtures.

FIXTURE INSTALLATIONS

With fixture installation, there are many rules and regulations to adhere to. Water supply is one issue. Access is another. Air gaps and overflows are factors. There are a host of requirements governing the installation of plumbing fixtures. We will start with the fixtures most likely to be found in residential homes, then look at those normally associated with commercial applications.

Residential Installation

Typical residential fixture installations can include everything from hose bibbs to bidets. This section will look at typical residential fixtures and tell you more about how they must be installed.

With most plumbing fixtures water enters and exits the fixture. Incoming water lines must be protected against freezing and backsiphonage. Freeze protection is usually accomplished through the placement of the piping. In cold climates it is advisable to avoid putting pipes in outside walls. Insulation is often applied to water lines to reduce the risk of freezing. Backsiphonage is typically avoided with the use of air gaps and backflow preventers. Shower valves and combination tub-shower valves are required to be balanced-pressure, thermostatic, or combination balanced-pressure/thermostatic valves. Mixing valves for tubs and showers must be set so that the maximum water temperature available at the device will not exceed 120°F. Temperature-actuated flow-reduction valves for individual fixtures are not to be used as a substitute for balanced-pressure, thermostatic, or combination shower-and-tub valves.

Some fixtures, such as lavatories and bathtubs, are equipped with overflow routes. These overflow paths must be designed and installed to prevent water from remaining in the overflow after the fixture is drained. They must also be installed in such a manner that backsiphonage cannot occur. This normally means nothing more than installing the faucet so that it is not submerged in water if the fixture floods. By keeping the faucet spout above the high-water mark, you have created an air gap. The path of a fixture's overflow must carry the overflowing water into the trap of the fixture. This should be done by integrating the overflow path with the same pipe that drains the fixture.

Bathtubs

Bathtubs must be equipped with wastes and overflows. Most codes require these wastes and overflows to have a minimum diameter of 1.5 inches. The method for blocking the waste opening must be approved. Common methods for holding water in a tub include the following:

- Plunger-style stoppers
- Rubber stoppers
- Lift-and-turn stoppers
- Push-and-pull stoppers.

Hand-held Shower

Some fixtures such as hand-held showers pose special problems. Since the shower is on a long hose, it could be dropped into a bathtub full of water. If a vacuum was formed in the water pipe while the shower head was submerged, the unsanitary water from the bathtub could be pulled back into the potable water supply. This is avoided with the use of an approved backflow preventer.

Fast Fact

Temperature-actuated flow-reduction valves for individual fixtures are not to be used as a substitute for balanced-pressure, thermostatic, or combination shower-and-tub valves.

Trade Tip

When a drainage connection is made with removable connections, such as slip nuts and washers, the connection must be accessible.

Drainage Connections

When a drainage connection is made with removable connections, such as slip nuts and washers, the connection must be accessible. If access is not practical, the connections must be soldered, solvent-cemented, or screwed to form a solid connection. This normally isn't a problem for sinks and lavatories, but it can create some difficulties with bathtubs. Many builders and homebuyers despise having an ugly access panel in the wall where their tub waste is located. To eliminate the need for this type of access, the tub waste can be connected with permanent joints. This could mean soldering a brass tub waste or gluing a plastic one. But if the tub waste is connected with slip nuts, an access panel is required.

Washing Machines

Washing machines generally receive their incoming water from boiler drains or laundry faucets. There is a high risk of a cross-connection when these devices are used with an automatic clothes washer. This type of connection must be protected against backsiphonage. The drainage from a washing machine must be handled by an indirect-waste receptor. An air break is required and is usually accomplished by placing the washer's discharge hose into a 2-inch pipe, such as a laundry tub, as an indirect waste receptor. The water supply to a bidet must also be protected against backsiphonage.

Dishwashers

Dishwashers are another likely source of backsiphonage. These appliances must be equipped with either a backflow protector or an air gap, installed on the water-supply piping. The drainage from dishwashers is handled differently in some codes.

It is common for the code to require the use of an air gap on the drainage of a dishwasher. These air gaps are normally mounted on the countertop or in the rim of the kitchen sink. The air gap forces the waste discharge of the dishwasher through open air and down a separate discharge hose. This eliminates the possibility of backsiphonage or a backup from the drainage system into the dishwasher. Some codes require dishwasher drainage to be separately trapped and vented or to be discharged indirectly into a properly trapped and vented fixture. Other codes allow the discharge hose from a dishwasher to enter the drainage system in several ways. It may be individually trapped. It may discharge into a trapped fixture. The discharge hose could be connected to a wye tailpiece in the kitchen-sink drainage. Or it may be connected to the waste connection provided on many garbage disposers.

Garbage Disposals

While we are on the subject of garbage disposers, be advised that they require a drain of at least 1.5 inches and must be trapped. It may seem to go without saying, but garbage disposers must have a water source. This doesn't mean you have to pipe a water supply to the disposer; a kitchen faucet provides adequate water supply to satisfy the code.

Floor Drains

Floor drains must have a minimum diameter of 2 inches. Remember, piping run under a floor may never be smaller than 2 inches in diameter. Floor drains must be trapped, usually must be vented, and must be equipped with removable strainers. It is necessary to install floor drains so that the removable strainer is readily accessible.

Laundry Trays

Laundry trays are required to have 1.5-inch drains. These drains should be equipped with crossbars or a strainer. Laundry trays may act as indirect-waste receptors for clothes washers. In the case of a multiple-bowl laundry tray, the use of a continuous waste is acceptable. All sinks are required to have a minimum drain diameter of 1.5 inches.

Lavatories

Lavatories are required to have drains of at least 1.25 inches in diameter. The drain must be equipped with some device to prevent foreign objects from entering the drain. These devices could include pop-up assemblies, crossbars, or strainers.

Showers

When installing a shower, it is necessary to secure the pipe serving the shower head with water. This riser is normally secured with a drop-ear elbow and screws. It is, however, acceptable to secure the pipe with a pipe clamp.

We are speaking only of showers, not tub-shower combinations. The frequent use of tub-shower combinations confuses many people. A shower has different requirements from those of a tub-shower combination. A shower drain must have a diameter of at least 1.5 inches. A drain with a minimum diameter of 2 inches is preferable. The reason for this is simple: In a tub-shower combination, a 1.5-inch drain is sufficient, because the walls of the bathtub will retain water until the smaller drain can remove it. A shower doesn't have high retaining walls, so a larger drain is preferable to clear the shower base of water more quickly. Shower drains must have removable strainers. The strainers should have a diameter of at least 3 inches, except when the shower is located in a bathtub.

Code may require all showers to contain a minimum of 900 square inches of shower base. This area must not be less than 30 inches in any direction. These measurements must be taken at the top of the threshold, and they must be interior measurements. The only objects allowed to protrude into this space are grab bars, faucets, and shower heads.

Showers are required to have a minimum height of 70 inches above the drain outlet. A shower advertised as a 30-inch shower may not meet code requirements. If the measurements are taken from the outside dimensions, the stall will not pass muster. There is one exception to the above ruling: Square showers with a rough-in of 32 inches may be allowed, but the exterior of the base may not measure less than 31.5 inches.

The waterproof wall enclosure of a shower or a tub-shower combination must extend from the finished floor to a height of no less than 6 feet. Another criterion for these enclosures is that they must extend at least 70 inches above the height of the drain opening. The enclosure walls must be at the higher of the two determining factors. An example of when this might come into play is a deck-mounted bathing unit. With a tub mounted in an elevated platform, an enclosure that extends 6 feet above the finished floor might not meet the requirement of being 70 inches above the drain opening.

Built-Up Shower Stalls

Though not as common as they once were, built-up shower stalls are still popular in high-end housing. These stalls typically use a concrete base covered with tile. You may never install one of these classic shower bases, but you need to know how just in case the need arises. These bases are often referred to as shower pans. Cement is poured into the pan to create a base for ceramic tile.

Before these pans can be formed, attention must be paid to the surface to be under the pan. The subfloor or other supporting surface must be smooth and able to accommodate the weight of the shower. When the substructure is satisfactory, you are ready to make your shower pan.

Shower pans must be made from a waterproof material. These pans used to be made of lead or copper. Today they are generally made with coated papers or vinyl materials. These flexible materials make the job much easier. When forming a shower pan, the edges of the pan material must extend at least 2 inches above the height of the threshold. The lining must not be nailed or perforated at any point less than 1 inch above the finished threshold. Local code may require the material to extend at least 3 inches above the threshold. The pan material must also be securely attached to the stud walls.

Some code requirements for shower regulations are more detailed. The shower threshold must be 1 inch lower than the other sides of the shower base, but the threshold must never be lower than 2 inches. The threshold must also never be higher than 9 inches. When installed for accessible facilities, the threshold may be eliminated.

Local codes may go on to require the shower base to slope toward the drain with a minimum pitch of 0.25 inch to the foot but not more than 0.5 inch per foot. The opening into the shower must be large enough to accept a shower door with minimum dimensions of 22 inches.

If PVC sheets are used as a shower base, they must have a minimum thickness of 0.040 inches. The sheets must be joined by solvent welding in accordance with the manufacturer's installation instructions.

Sheet lead used for pan material is required to weigh not less than 4 pounds per square foot. The lead is to be coated with an asphalt paint or other approved coating. Lead sheets must be insulated from conducting substances other than the connecting drain by 15-pound asphalt felt or its equivalent. Sheet lead must be joined by burning.

Sheet copper can also be used as a shower-pan liner. The copper must not weigh less than 12 ounces per square foot. Insulation must protect the copper from conducting substances other than the connecting drain by 15-pound asphalt felt or its equivalent. When sheet copper is joined, it must be brazed or soldered.

The drains for this type of shower base are new to many young plumbers. Plumbers who have worked under my supervision have attempted to use standard shower drains for these types of bases. You cannot do that—at least, not if you don't want the pan to leak. This type of shower base requires a drain that is similar to some floor drains.

The drain must be installed in a way that will not allow water that might collect in the pan to seep around the drain and down the exterior of the pipe. Any water entering the pan must go down the drain. A proper drain will have a flange that sits beneath the pan material. The pan material is cut to allow water into the drain. Then another part of the drain is placed over the pan material and bolted to the bottom flange. The compression of the top piece and the bottom flange, with the pan material wedged between them, will create a watertight seal. Then the strainer portion of the drain will screw into the bottom flange housing. Since the strainer is on a threaded extension, it can be screwed up or down to accommodate the level of the finished shower pan.

Fast Fact

All plumbing faucets and valves using both hot and cold water must be piped in a uniform manner in which the hot water is piped to the left side of the faucet or valve.

Sinks

Sinks are required to have drains with a minimum diameter of 1.5 inches. Strainers or crossbars are required in the sink drain. If you look, you will see that with basket strainers the basket part serves as a strainer, with crossbars below it. This provides protection from foreign objects, even when the basket is removed. If a sink is equipped with a garbage disposer, the drain opening in the sink should have a diameter of at least 3.5 inches.

Toilets

Toilets are usually required to be water-saver models. The older models, which use 5 gallons per flush, are no longer allowed in many jurisdictions for new installations.

The seat on a residential water closet must be smooth and sized for the type of water closet it is serving. This usually means that the seat will have a round front.

The fill valve or ballcock for toilets must be of the antisiphon variety. A fill-valve backflow preventer must be located at least 1 inch above the full opening of the overflow pipe. Older ballcocks are still sold that are not of the antisiphon style. Just because these units are available doesn't make them acceptable. Don't use them—you will be putting your license and yourself on the line.

Toilets of the flush-tank type are required to be equipped with overflow tubes. These overflow tubes do double duty as refill conduits. The overflow tube must be large enough to accommodate the maximum water intake entering the water closet at any given time. All parts in a flush tank must be accessible for repair and replacement. The overflow tube must discharge into the water closet that it is connected to. An overflow pipe must be installed so that the opening of the pipe is located above the flood-level rim of the water closet or above a secondary overflow in the flush tank.

Whirlpool Tubs

Whirlpool tubs must be installed as recommended by the manufacturer. All whirlpool tubs must be installed to allow access to the unit's pump. The pump's drain should be pitched to allow the pump to empty its volume of water when the whirlpool is drained. The whirlpool pump should be positioned above the fixture's trap.

Faucets and Valves

All plumbing faucets and valves using both hot and cold water must be piped in a uniform manner in which the hot water is piped to the left side of the faucet or valve. Cold water should be piped to the right side of the faucet or valve. This uniformity reduces the risk of unwarranted burns from hot water.

Valves or faucets used for showers must be designed to provide protection from scalding. This means that any valve or faucet used in a shower must be pressure-balanced or contain a thermostatic mixing valve. The temperature control must not allow the water temperature to exceed 110°F in some regions and 120°F in others. This provides safety, especially to the elderly and the very young, against scalding injuries from the shower. Not all codes require these temperature-controlled valves in residential dwellings. The thermostatic mixing valves must be sized according to the peak demand of fixtures located downstream of the valve. A water-heater thermostat cannot be used as the temperature-control device for compliance on this issue.

COMMERCIAL APPLICATIONS

Drinking fountains are a common fixture in commercial applications. Restaurants use garbage disposers that are so big that it can take two plumbers to move them. Gang showers are not uncommon in school gyms and health clubs. Urinals are other common commercial fixtures. Water closets installed for commercial applications often differ from residential toilets. Special fixtures and applications exist for some unusual plumbing fixtures, such as baptismal pools in churches. This section is going to take you into the commercial field and show you how plumbing needs vary from residential uses to commercial applications.

Let's start with drinking fountains and water coolers. The main fact to remember about water coolers and fountains is that they are not allowed in toilet facilities. You may not install a water fountain in a room that contains a water closet. If the building for which a plumbing diagram is being designed will serve water, such as a restaurant, or if the building will provide access to bottled water, drinking fountains and water coolers may not be required.

Commercial garbage disposers can be big. These monster grinding machines require a drain with a diameter of no less than 2 inches. Commercial disposers must have their own drainage piping and trap. As with residential disposers, commercial disposers must have a cold-water source. In some jurisdictions, the water source must be of an automatic type. These large disposers may not be connected to a grease interceptor. Commercial dishwashing machines must discharge through an air gap or air break into a standpipe or waste receptor as called for in the local plumbing code.

Garbage-can washers are not fixtures you will find in the average home, but they are not uncommon in commercial applications. Due to the nature of this fixture, the water supply to the fixture must be protected against backsiphonage. This can be done with either a backflow preventer or an air gap. The waste pipe from these fixtures must have an individual trap. The receptor that collects the residue from the garbage-can washer must be equipped with a removable strainer, capable of preventing the entrance of large particles into the sanitary drainage system.

Showers for commercial or public use can be very different from those found in a residence. It is not unusual for showers in commercial-grade plumbing to be gang showers. This amounts to one large shower enclosure with many showerheads and shower valves. In gang showers, the shower floor must be properly graded toward the shower drain or drains. The floor must be graded in a way to prevent water generated at one shower station from passing through the floor area of another shower station.

The methods employed to divert water from each shower station to a drain are up to the designer, but it is imperative that water used by one occupant not pass into another bather's space.

Fast Fact

The main fact to remember about water coolers and fountains is that they are not allowed in toilet facilities.

Urinals are not a common household item, but they are typical fixtures in public toilet facilities. The amount of water used by a urinal in a single flush should be limited to a maximum of 1.5 gallons. Water supplies to urinals must be protected from backflow. Only one urinal may be flushed by a single flush valve. When urinals are used, they must not take the place of more than half the water closets normally required. Public-use urinals are required to have a water-trap seal that is visible and unobstructed by strainers.

Floor and wall conditions around urinals are another factor to be considered. These areas are required to be waterproof and smooth. They must be easy to clean, and they may not be made from an absorbent material. Waterproof materials must extend to at least 1 foot on each side of the urinal. This measurement is taken from the outside edge of the fixture. The material is required to extend from the finished floor to a point 4 feet away. The floor under a urinal must be made of this same type of material, and the material must extend to a point at least 1 foot in front of the furthest portion of the urinal.

Commercial-grade water closets can vary from residential requirements. The toilets used in public facilities must have elongated bowls. These bowls must be equipped with elongated seats. Further, the seats must be hinged, and they must have open or split fronts.

Flush valves are used almost exclusively with commercial-grade fixtures. They are used on water closets, urinals, and some special sinks. If a fixture depends on trap siphonage to empty itself, it must be equipped with a flush valve or a properly rated flush tank. These valves or tanks are required for each fixture in use.

Flush valves must be equipped with accessible vacuum breakers. Flush valves must be rated as water-conserving. These valves must be able to regulate water pressure, and they must open and close fully. If water pressure is not sufficient to operate a flush valve, other measures, such as a flush tank, must be incorporated into the design. All manually operated flush tanks should be controlled by an automatic filler, designed to refill the tank after each use. The automatic filler must be equipped to cut itself off when the trap seal is replenished and the flush tank is full. If a flush tank is designed to flush automatically, the filler device should be controlled by a timer.

Fast Fact

Flush valves must be equipped with accessible vacuum breakers.

SPECIAL FIXTURES

Special fixtures are just that—special. Fixtures that might fall into this category include church baptismal pools, swimming pools, fishponds, and other such arrangements. The water pipes to any of these special fixtures must be protected against backsiphonage.

There is an entire group of special fixtures that are normally found only in facilities providing healthcare. The requirements for these fixtures are extensive. While you may never work with these specialized fixtures, you should know their code requirements. This section will provide you with the information you may need.

Many healthcare fixtures are required to be made to a higher standard of materials. They may be required to endure excessive heat or cold. Many of these special fixtures must be protected against backflow. The prevention of backflow extends to the drainage system as well as the potable water supply. All special fixtures must be of an approved type.

Sterilizers

Any concealed piping that serves special fixtures and that may require maintenance or inspection must be accessible. All piping for sterilizers must be accessible. Steam piping to a sterilizer should be installed with a gravity system to control condensation and to prevent moisture from entering the sterilizer. Sterilizers must be equipped with a means to control steam vapors. The drains from sterilizers must be piped as indirect wastes. Sterilizers are required to have leak detectors. These leak detectors are designed to expose leaks and to carry unsterile water away from the sterilizer. The interiors of sterilizers may not be cleaned with acid or other chemical solutions while the sterilizers are connected to the plumbing system.

Fast Fact

All piping for sterilizers must be accessible.

Clinical Sinks

Clinical sinks are sometimes called bedpan washers. Clinical sinks are required to have an integral trap. The trap seal must be visible, and the contents of the sink must be removed by siphonic or blowout action. The trap seal must be automatically replenished, and the sides of the fixture must be cleaned by a flush rim at every flushing of the sink. These special fixtures are required to connect to the DWV system in the same manner as a water closet. When clinical sinks are installed in utility rooms, they are not meant to be a substitute for a service sink. On the other hand, service sinks may never be used to replace a clinical sink. Devices for making or storing ice cannot be placed in a soiled utility room.

Fluid-Suction Systems

Vacuum-system receptacles are to be built into cabinets or cavities, but they must be visible and readily accessible. Bottle-suction systems used for collecting blood and other human fluids must be equipped with overflow-prevention devices at each vacuum receptacle. Secondary safety receptacles are recommended as an additional safeguard. Central fluid-suction systems must provide continuous service. If a central suction system requires periodic cleaning or maintenance, it must be installed so that it can continue to operate, even while cleaning or maintenance is being performed. When central systems are installed in hospitals, they must be connected to

Trade Tip

Clinical sinks are sometimes called bedpan washers.

emergency power facilities. The vent discharge from these systems must be piped separately to the outside air above the roof of the building.

Waste originating in a fluid-suction system that is to be drained into the normal drainage piping must be piped into the drainage system with a direct-connect, trapped arrangement; indirect-waste connections of this type of unit are not allowed. Piping for these fluid-suction systems must be noncorrosive and have a smooth interior surface. The main pipe must have a diameter of no less than 1 inch. Branch pipes must not be smaller than 0.5 inch. All piping is required to have accessible cleanouts and must be sized according to manufacturer's recommendations. The air flow in a central fluid-suction system should not be allowed to exceed 5000 feet per minute.

Special Vents

Institutional plumbing uses different styles of vents for some equipment than what is encountered with normal plumbing. One such vent is called a local vent. One example of use for a local vent pertains to bedpan washers. A bedpan washer must be connected to at least one vent, with a minimum diameter of 2 inches, and that vent must extend to the outside air above the roof of the building.

These local vents are used to vent odors and vapors. Local vents may not tie in with vents from the sanitary plumbing or sterilizer equipment. In multistory buildings a local vent stack may be used to collect the discharge from individual local vents for multiple bedpan washers located above each other. A 2-inch stack can accept up to three bedpan washers. A 3-inch stack can handle six units, and a 4-inch stack will accommodate up to twelve bedpan washers. These local vent stacks are meant to tie into the sanitary drainage system, and they must be vented and trapped if they serve more than one fixture.

Fast Fact

Waste originating in a fluid suction system that is to be drained into the normal drainage piping must be piped into the drainage system with a direct-connect, trapped arrangement; indirect-waste connections for this type of unit are not allowed.

Each local vent must receive water to maintain its trap seal. The water source comes from the water supply for the bedpan washer served by the local vent. A minimum of 0.25-inch tubing should be run to the local vent, and it should discharge water into the vent each time the bedpan washer is flushed.

Vents serving multiple sterilizers must be connected with inverted wye fittings, and all connections must be accessible. Sterilizer vents are intended to drain to an indirect waste. The minimum diameter of a vent for a bedpan sterilizer is 1.5 inches. When serving a utensil sterilizer, the minimum vent size is 2 inches. Vents for pressure-type sterilizers must be at least 2.5 inches in diameter. When serving a pressure instrument sterilizer, a vent stack must be at least 2 inches in diameter. Up to two sterilizers of this type may be on a 2-inch vent. A 3-inch stack can handle four units.

Water Supply

Hospitals are required to have at least two water services. These two water services may, however, connect to a single water main. Hot water must be made available to all fixtures, as required by the fixture manufacturer. All water heaters and storage tanks must be of a type approved for the intended use.

Some jurisdictions require the hot-water system to be capable of delivering 6.5 gallons of 125° water per hour for each bed in a hospital. Some codes further require hospital kitchens to have a hot-water supply of 180° water equal to four gallons per hour for each bed. Laundry rooms are required to have a supply of 180° water at a rate of 4.5 gallons per hour for each bed. Local code may require hot-water storage tanks to have capacities equal to no less than 8 percent of the water-heating capacity. Some codes dictate the use of copper in submerged steam-heating coils. If a building is higher than

Fast Fact

Local vents may not tie in with vents from the sanitary plumbing or sterilizer equipment.

Trade Tip

Hospitals are required to have at least two water services.

three levels, the hot-water system must be equipped to circulate. Valves are required on the water-distribution piping to fixture groups.

Backflow Prevention

When backflow-prevention devices are installed, they must be installed at least 6 inches above the flood-level rim of the fixture. In the case of handheld showers, the height of installation shall be 6 inches above the highest point that the hose can be used.

In most cases, hospital fixtures will be protected against backflow by the use of vacuum breakers. However, a boiling-type sterilizer should be protected with an air gap. Vacuum suction systems may be protected with either an air gap or a vacuum breaker.

This has been a long chapter, but it was necessary to give you all the pertinent details on fixtures. As you now know, fixtures are not as simple as they may first appear. There are numerous regulations to learn and apply when installing plumbing fixtures. Your local jurisdiction may require additional code-compliance information. As always, check with your local authority before installing plumbing.

CHAPTER

5

Water Heaters

Water heaters are only a small portion of a plumbing system, but they are important and they can be dangerous. Many plumbers take water heaters and the codes pertaining to them lightly. This is a mistake. Water heaters can become lethal if they are not installed properly. I feel very strongly about the safety issues associated with water heaters. My position is based largely on decades of watching plumbers and plumbing contractors take shortcuts with water heaters that could be disastrous.

GENERAL PROVISIONS

Water heaters are sometimes used as a part of a space-heating system. When this is the case, the maximum outlet water temperature for the water heater is 140°F unless a tempering valve is used to maintain an acceptable temperature in the potable water system. It is essential that all potable water in the water heater be maintained throughout the entire system. Potability of water must be maintained at all times. Every water heater is required to be equipped with a drain valve near the bottom of the water heater. This is true, too, for hot-water storage tanks. All drain valves must conform to ASSE 1005.

Fast Fact

The location of water heaters and hot-water storage tanks is important. Code requires both water heaters and hot-water storage tanks to be accessible for observation, maintenance, servicing, and replacement.

The location of water heaters and hot-water storage tanks is important. Code requires both water heaters and hot-water storage tanks to be accessible for observation, maintenance, servicing, and replacement. Every water heater is required to bear a label of an approved agency.

The temperature of water delivered from a tankless water heater may not exceed 140°F when used for domestic purposes. This portion of the code does not supersede the requirement for protective shower valves, as detailed in the code.

All water heaters must be third-party certified. Water heaters must be installed in accordance with manufacturer's requirements. Oil-fired water heaters must be installed in conformance with the requirements of the ICC Electrical Code. Gas-fired water heaters are required to meet the criteria of the International Fuel Gas Code.

All storage tanks and water heaters installed for domestic hot water must have the maximum allowable working pressure clearly and indelibly stamped in the metal or marked on a plate welded thereto or otherwise permanently attached. All markings of this type must be in accessible positions outside the tanks. Inspection or reinspection of these markings must be easily performed.

Trade Tip

Every hot-water supply system is required to be fitted with an automatic temperature control.

Fast Fact

When earthquake loads are applicable, water-heater supports must be designed and installed for seismic forces in accordance with the International Building Code.

Every hot-water supply system is required to be fitted with an automatic temperature control. The control must be capable of being adjusted from a minimum temperature to the highest acceptable setting for the intended operating range.

INSTALLATION

All water heaters are required to be installed in accordance with the manufacturer's recommendations and the plumbing code. Water heaters that are fueled by gas or oil must conform to both the plumbing code and the mechanical or gas code. Electric water heaters must conform to the requirements of the plumbing code and the provisions of NFPA 70, as listed in the plumbing code.

When earthquake loads are applicable, water-heater supports must be designed and installed for seismic forces in accordance with the International Building Code.

When water heaters are installed in attics, special provisions must be made. For example, the attic must be provided with an opening and unobstructed passageway large enough to allow for the removal of the water heater. This should be common sense, but it is also part of the plumbing code. There are many measurements that come into play when planning the exit route for an attic water heater. They are as follows:

- Minimum height: 30 inches
- Minimum width: 22 inches
- Maximum length: 20 feet.

A continuous solid floor is required in the exit area, and the flooring must be at least 24 inches wide. Another requirement calls for a level service area with minimum dimensions of 30 inches deep and 30 inches wide. This service area must be made in front of the water heater or wherever the service area of the water heater is located. A clear access opening with minimum dimensions of 20 inches by 30 inches—large enough to allow removal of the water heater—is needed.

CONNECTIONS

Making connections to water heaters is not difficult, but the manner in which the connections are made must conform to code requirements. The first consideration is the installation of cutoff valves. A cold-water branch line from a main water supply to a hot-water storage tank or water heater must be provided with a cutoff valve that is accessible on the same floor, located near the equipment, and serving only the hot-water storage tank or water heater. The valve used must not interfere with or cause a disruption of the cold-water supply to the remainder of the cold-water system.

Any means of connecting a circulating water heater to a tank must provide for proper circulation of water through the water heater. All piping that is required for the installation of appliances that will draw from the water heater or storage tank must comply with all provisions of the plumbing and mechanical codes.

SAFETY

Compliance with safety requirements is essential when installing or replacing a water heater or hot-water storage tank. One major concern is the si-

Trade Tip

Every electric water heater must be provided with its own electrical disconnect switch in close proximity to the water heater. In the case of gas-fired or oil-fired water heaters, cutoff valves must be installed close to the water heaters to stop the fuel flow when needed.

Fast Fact

Energy-cutoff valves are required on all water heaters that are automatically controlled.

phoning of water from a water heater or storage tank. An antisiphon device of an approved type is required to prevent siphoning. A cold-water dip tube with a hole at the top or a vacuum-relief valve installed in the cold-water supply line above the top of the water heater or storage tank is an acceptable means of protection. Some water heaters and storage tanks receive their incoming water from the bottom of the unit. These types of heaters and tanks must be supplied with an approved vacuum-relief valve that complies with ANSI Z21.22.

Energy-cutoff valves are required on all water heaters that are automatically controlled. The energy-cutoff valve is designed to cut off the supply of heat energy to the water tank before the temperature of the water in the tank exceeds 210°F. The installation of an energy-cutoff valve does not remove the need for a temperature-and-pressure relief valve—both are required.

Every electric water heater must be provided with its own electrical disconnect switch that is in close proximity to the water heater. In the case of gas-fired or oil-fired water heaters, cutoff valves must be installed close to the water heaters to stop the fuel flow when needed.

RELIEF VALVES

Pressure-relief valves and temperature-relief valves or combination temperature-and-pressure relief valves (the most commonly used) are required on all water heaters and storage tanks that are operating above atmospheric pressure. The valves used must be approved and conform to ANSI Z21.22 ratings. Relief valves must be of a self-closing (levered) type. In no case shall the relief valve be used as a means of controlling thermal expansion.

Trade Tip

Never omit the installation of required relief valves. Failure to do so can be catastrophic.

Relief valves must be installed in the shell of a water-heater tank. Any temperature-relief valve must be installed so that it is actuated by the water in the top 6 inches of the tank being served by the valve. When separate tanks are used, the valves must be installed on the tank and not between the water heater and the storage tank. It is prohibited to install a cutoff valve or check valve between a relief valve and the water heater or tank being serviced by the relief valve. Never omit the installation of required relief valves. Failure to do so can be catastrophic.

All relief valves, whether temperature, pressure, or a combination of the two, and all energy-cutoff devices must bear a label of an approved agency. The valves and devices must have a temperature setting of not more than 210°F and a pressure setting that does not exceed the tank or water-heater manufacturer's rated working pressure or 150 psi, whichever is less. The relieving capacity of each relief valve must equal or exceed the heat input to the water heater or storage tank.

Since relief valves may create a discharge, the disposal must be dealt with. In no case is it allowable for the discharge tube from a relief valve to be connected directly to a drainage system. The discharge tube must be provided in a full-size tube or pipe that is the same size as the outlet of the relief valve. You have two choices on the termination point of the discharge tube. It can be piped to the outside of a building or it can terminate over an indirect waste receptor that is located inside a building.

When freezing conditions may exist, the discharge tubing or piping for a relief valve must be protected. This is done by having the tubing discharge through an air gap and into an indirect waste receptor that is located in a heated space. Local regulations may allow some other form of installation, so check your local code requirements.

Any risk of personal injury or property damage must be avoided when piping a discharge tube from a relief valve. The discharge piping must be installed so that it is readily observable by building occupants. Traps on discharge tubes and pipes are prohibited. All discharge piping must drain by gravity. The tubing must terminate atmospherically not more than 6 inches above the floor, and the end of the discharge tubing or piping is not allowed to be threaded. When discharge piping is installed so that it exits a room or enclosure housing a water heater and a relief valve that discharges into an indirect waste receptor, there must be an air gap installed before or at the point of exit. Discharge tubes from relief valves must not discharge into a safety pan. Materials used for discharge piping must be made according to the standards listed in the plumbing code and tested, rated, and approved for such use in accordance with ASME A112.4.1.

Safety pans are required for water heaters and storage tanks that are installed in locations where leakage may cause property damage. Water heaters and storage tanks must be placed in safety pans that are constructed of galvanized steel or other approved metal materials. The minimum thickness of the metallic pan must be 24-gauge. Electric water heaters must be installed in pans when leakage may cause property damage, and the pan must be made of 24-gauge metal or a high-impact plastic that has a minimum thickness of 0.0625 inch. All piping from safety-pan drains must be made with materials approved by the plumbing code.

Safety pans must have a minimum depth of 1.5 inches and be of sufficient size and shape to receive all dripping or condensate from the tank or water heater contained in the pan. A safety pan must drain by an indirect waste. The drainage pipe or tube from the pan must have a minimum diameter of 1 inch or the outlet diameter of the relief valve, whichever is larger.

Fast Fact

Safety pans are required for water heaters and storage tanks that are installed in locations where leakage may cause property damage.

The drain tube or pipe from a safety pan must run full-size for its entire length and either terminate over a suitably located indirect waste receptor or floor drain or extend to the exterior of the building and terminate not less than 6 inches or more than 24 inches above the adjacent ground level.

Unfired hot-water storage tanks must be insulated so that heat loss is limited to a maximum of 15 British thermal units (Btu) per hour per square foot of external tank surface area. For purposes of determining this heat loss, the design ambient temperature cannot be higher than 65°F.

VENTING

The venting of water heaters that require venting is regulated by the plumbing code. All venting materials used must be in compliance with all code requirements. Venting systems might consist of approved chimneys, Type B vents, Type L vents, or plastic pipe. The recommendations of the equipment manufacturer must be observed in selecting the proper venting material and installation procedure.

Vents must be designed and installed to develop a positive flow adequate to convey all products of combustion to the outside atmosphere. Condensing appliances that cool flue gases nearly to the dew point within the appliance, resulting in low vent-gas temperatures, may use plastic venting materials and vent configurations that are unsuitable for noncondensing appliances. All unused openings in a venting system must be closed or capped to the satisfaction of the local code-enforcement officer.

Type B vents are not allowed for use with water heaters that are converted readily to the use of solid or liquid fuels. Water heaters listed for use with chimneys only may not be vented with Type B vents. Manually operated dampers must not be installed in chimneys, vents, or chimney or vent connectors of fuel-burning water heaters. Fixed baffles on the water-heater side of draft hoods and draft regulators are not to be considered as dampers.

Connectors

Vent connectors used for gas water heaters with draft hoods may be constructed of noncombustible materials with corrosion resistance not less than that of galvanized sheet steel and of a thickness not less than that specified in the code. Or they may be of Type B or Type L vent material. When single-wall metal vent connectors are used, they must be securely supported and joints fastened with sheet-metal screws, rivets, or another approved means. Such connectors must not originate in an unoccupied attic or concealed space; must not pass through any attic, inside wall, floor, or concealed space; and must be located in the same room or space as the fuel-burning water heater.

Support

Combustion products, vents, vent connectors, exhaust ducts from ventilating hoods, chimneys, and chimney connectors must not extend into or through any air duct or plenum, except when the venting system passes through a combustion air duct. The base of all vents supported from the ground must rest on a solid masonry or concrete base extending at least 2 inches above adjoining ground level. If the base of a vent is not supported from the ground and is not self-supporting, it must rest on a firm metal or masonry support. All venting systems must be supported adequately for their weight and design. No water heater is allowed to be vented into a fireplace or into a chimney that serves a fireplace.

Offsets

With minor exceptions, gravity vents must extend in a generally vertical direction with offsets not exceeding 45 degrees. These vents are allowed to have one horizontal offset of not more than 60 degrees. All offsets must be supported properly for their weight and must be installed to maintain proper clearance to prevent physical damage and the separation of joints.

Offsets with angles of more than 45 degrees are considered to be horizontal offsets. Horizontal vent connectors must not be greater than 75 percent of the vertical height of the vent and must comply with all code regulations.

Vent connectors in a gravity-type venting system must have a continuous rise of not less than 0.25 inch per foot of developed length, measured from the appliance vent collar to the vent. If a single-wall metal vent connector is allowed and installed, it must have a minimum clearance of 6 inches from any combustible material.

Termination

Vents must terminate above the roof surface of the building being served. The pipe must pass through a flashing and terminate in an approved or listed vent cap that is installed in accordance with the manufacturer's recommendations. There is an exception: A direct vent or mechanical draft appliance is acceptable when installed according to its listing and the manufacturer's instructions.

Gravity-type venting systems, with the exception of venting systems that are integral parts of a listed water heater, must terminate at least 5 feet above the highest vent collar being served. Type B gas vents with listed vent caps 12 inches in size or smaller can be terminated in accordance with the code requirements so long as they are at least 8 feet from a vertical wall or similar obstruction. All other Type B vents must terminate not less than 2 feet above the highest point where they pass through the roof and at least 2 feet higher than any portion of a building within 10 feet.

Type L vents must not terminate less than 2 feet above the roof through which they pass, nor less than 4 feet from any portion of the building that extends at an angle of more than 45 degrees upward from the horizontal. No vent system is allowed to terminate less than 4 feet below, 4 feet horizontally from, or 1 foot above any door, openable window, or gravity air inlet into any building. As usual, there are exceptions.

Terminals of direct vent appliances with inputs of 50,000 Btu/hour or less can be located at least 9 inches from an opening through which combustion products could enter a building. Appliances with inputs in excess of 50,000 Btu/hour but not exceeding 65,000 Btu/hour require 12-inch vent-termination clearances. The bottom of the vent terminal and the air intake must be located at least 12 inches above grade.

Area

The internal cross-sectional area of a venting system must not be less than the area of the vent collar on the water heater, unless the venting system has been designed in accordance with other code requirements. In no case shall the area be less than 7 square inches, unless the venting system is an integral part of a listed water heater.

Multiple Appliances

It is acceptable to connect multiple oil- or listed gas-burning appliances to a common gravity-type venting system, provided the appliances are equipped with an approved primary safety control capable of shutting off the burners and the venting system is designed in compliance with the code requirements.

Multiple appliances connected to a common vent system must be located within the same story of the building, unless an engineered system is being used. The inlets for multiple connections must be offset in such a way that no inlet is opposite another inlet. Oval vents may be used for multiple appliance venting, but the venting system must be not less than the area of the largest vent connector plus 50 percent of the areas of the additional vent connectors.

Existing Systems

New water heaters installed as replacements must meet code criteria before they can be connected to existing venting systems. The existing system must have been installed lawfully at the time of its installation. Code compliance with the internal area of the venting system is a must. Any connection must be made in a safe manner.

Draft Hoods

Draft hoods for water heaters must be located in the same room or space as the combustion air opening for the water heater. The draft hood must be installed in the position for which it is designed and must be located so that the relief opening is not less than 6 inches from any surface other than the water heater being served, measured in a direction 90 degrees to the plane of the relief opening. Exceptions could exist if a manufacturer's recommendations vary.

Masonry Chimneys

Existing masonry chimneys with not more than one side exposed to the outside can be used to vent a gas water heater. There are, however, some conditions that apply. The local code may require unlined chimneys to be lined with approved materials. The effective cross-sectional area of the chimney must not be more than four times the cross-sectional area of the vent and chimney connectors entering the chimney. The effective area of the chimney when connected to multiple connectors must not be less than the area of the largest vent or connector plus 50 percent of the area of the additional vent or connector.

Automatically controlled gas water heaters connected to a chimney that also serves equipment burning liquid fuel must be equipped with an automatic pilot. A gas water-heater connector and a connector from an appliance burning liquid fuel may be connected to the same chimney through separate openings, providing the gas water heater is vented above the other fuel-burning appliance; or both can be connected through a single opening if joined by a suitable fitting located at the chimney. Multiple connections must not be made at the same horizontal plane of another inlet. Any chimney used must be clear of obstructions and cleaned if previously used for venting solid or liquid fuel-burning appliances.

Chimney Connectors

Chimney connectors must comply with code requirements as set forth in tables in the local code book. When multiple connections are made, the connector, the manifold, and the chimney must be sized properly. Gravity vents must not be connected to vent systems served by power venters unless the connection is made on the negative-pressure side of the power exhauster. Single-wall metal chimneys require a minimum clearance of 6 inches from combustible materials.

Connectors must be kept as short and as straight as possible. Water heaters are required to be installed as close as possible to the venting system. Connectors must not be longer than 75 percent of the portion of the venting system above the inlet connection unless they are part of an approved engineered system.

A connector to a masonry chimney must extend through the wall to the inner face of the liner but not beyond. The connector must be cemented to the masonry. A thimble may be used to facilitate removal of the connector for cleaning, in which case the thimble must be permanently cemented in place. Connectors cannot pass through any floor or ceiling.

Trade Tip

It is acceptable to vent water heaters with mechanical draft systems of either forced- or induced-draft design.

Draft regulators are required in connectors serving liquid fuel-burning water heaters, unless the water heater is approved for use without a draft regulator. When used, draft regulators must be installed in the same room or enclosure as the water heater in such a manner that there is no difference in pressure between air in the vicinity of the regulator and the combustion air supply.

Mechanical Draft Systems

It is acceptable to vent water heaters with mechanical draft systems of either forced- or induced-draft design. Forced-draft systems must be designed and installed to be gastight or to prevent leakage of combustion products into a building. Connectors vented by natural draft must not be connected to mechanical draft systems operating under positive pressure. Systems using a mechanical draft system must be designed to prevent the flow of gas to the main burners when the draft system is not performing so as to satisfy the operating requirements of the water heater for safe performance.

Exit terminals of mechanical draft systems must be located not less than 12 inches from any opening through which combustion products could enter the building, nor less than 7 feet above grade when located adjacent to public walkways.

Ventilating Hoods

Ventilating hoods can be used to vent gas-burning water heaters installed in commercial applications. Dampers are not allowed when automatically operated water heaters are vented through natural draft ventilating hoods. If a power venter is used, the water-heater control system must be interlocked so that the water heater will operate only when the power venter is in operation.

COMMENTARY

I've been plumbing for over 25 years. During these years I've done just about every type of plumbing that there is. My career has involved working with all sorts of plumbers and plumbing contractors. As a plumbing contractor for the last 20 years, I've used a lot of other plumbing contractors as subcontractors. My work has even extended into teaching code classes and apprenticeship classes at Central Maine Technical College. I'm telling you this to give you an idea of my background and experience in the industry. This is because I want to tell you something that I feel is extremely important about water heaters.

During my time in the field, I've run into many occasions when plumbing contractors failed to obtain permits for the installation of water heaters. A plumbing permit is required for every water-heater installation and replacement. Yet a good number of contractors feel that they can get by without a permit. I've heard dozens of contractors say that it takes more time to get a permit and an inspection than it does to install a water heater. This can be true, but the point is that a permit and an inspection are required by code.

Some code offices are more active than others in enforcing the need for permits and inspections when water heaters are being replaced. But all codes that I know of do require a permit and an inspection for the installation or replacement of a water heater. This shouldn't be a big deal, but it is, and it can be a very big deal.

A few years ago I lost the service contract on more than 180 apartment units because I refused to replace a water heater without a permit and an inspection. I didn't like losing the account, but I was not going to violate the plumbing code and put myself and my business at risk for a lawsuit for any job. It can be hard for reputable plumbing contractors to compete with bootleggers who are willing to work without permits. The time and money spent on permits and inspections do drive up the cost of a job. Still, you should never install or replace a water heater without adhering to the plumbing code.

Why do I feel so strongly about permits for water heaters? There are several reasons. First, code requires a permit and an inspection, so this should be reason enough. Secondly, installing a water heater illegally can open up a huge risk for a lawsuit. If you install a water heater in accordance with code requirements and a problem erupts later, your insurance should cover your losses. This probably would not be the case if you had installed the water heater in violation of the plumbing code. Leaving ethics, morals, and lawsuits out of it, there is still the issue of personal safety.

Believe it or not, I have found water heaters where a plug had been installed in place of a relief valve. Any plumber with even minimal experience knows the risk involved. I've found cutoff valves installed in illegal locations around water heaters. If the proper safety precautions are not taken, a water heater can become a large bomb. The explosions can be very violent.

I won't try to beat you over the head with my vision, but seriously consider the risk you will be taking if you don't install water heaters properly and in accordance with all code provisions. Protect yourself by getting permits and inspections. With this said, let's move to the next chapter and explore water supply and distribution.

CHAPTER

6

Water Supply and Distribution

Potable water is essential to life. Water is often taken for granted in today's society, but it shouldn't be. Without good water, we would all die. Part of a plumber's responsibility is to provide safe drinking water. The requirements for ensuring that water is safe to drink are many. This chapter deals with the installation of potable water systems. It details approved types of piping and installation methods. These guidelines must be followed to ensure the health of everyone.

Potable water is water that is safe for drinking, cooking, and bathing, among other uses. Any building intended for human occupancy that has plumbing fixtures installed must have a potable water supply. If a building is intended for year-round habitation or a place of work, both hot and cold potable water must be made available. All fixtures that are intended for use in bathing, drinking, cooking, food processing, or medical products must have potable water available, and only potable water. It is permissible to use nonpotable water for flushing toilets and urinals. Now that you know what must be equipped with potable water, let's see how to get the water to the fixtures.

Fast Fact

A water-service pipe must have a diameter of at least 0.75 inch.

WATER MAIN

The main water pipe delivering potable water to a building is called a water service. A water-service pipe must have a diameter of at least 0.75 inch. The pipe must be sized, according to code requirements, to provide adequate water volume and pressure to the fixtures.

Ideally, a water-service pipe should be run in a private trench from the primary water source to the building. By a private trench I mean a trench not used for any purpose except water service. However, it is normally allowable to place the water service in a trench used by a sewer or building drain so long as specific installation requirements are followed. The water-service pipe must be separated from the drainage pipe. The bottom of the water-service pipe may not be closer than 12 inches to the drainage pipe at any point.

A shelf must be made in the trench to support the water service. The shelf must be made solid and stable, at least 12 inches above the drainage pipe. If the water service is not located above the sewer, there must be at least 5 feet of undisturbed or compacted earth between the two pipes. It is not acceptable to have a water service located in an area where pollution is probable. A water service should never run through, above, or under a waste-disposal system, such as a septic field. It is unacceptable to install a water-service pipe in soil or ground water that is contaminated with solvents, fuels, organic compounds, or other detrimental materials that may cause permeation,

Trade Tip

A water service should never run through, above, or under a waste-disposal system, such as a septic field.

CODE UPDATE

Access must be provided to manifolds with integral factory or field-installed valves.

corrosion, degradation, or structural failure of the piping material. If such conditions are suspected to be present, a chemical analysis of the soil or ground water is required. Any materials used for a water service must be approved for the intended use, application, and location.

The required separation distance between a water service and a sewer does not apply when the water service crosses a sewer pipe and is sleeved to at least 5 feet horizontally from the sewer pipe centerline on both sides of the crossing.

If a water service is installed in an area subject to flooding, the pipe must be protected. Water services must also be protected against freezing. The depth of the water service will depend on the climate of the location. Check with your local code officer to see how deep a water service pipe must be buried in your area. Care must be taken when backfilling a water-service trench. The backfill material must be free of objects, such as sharp rocks, that may damage the pipe.

When a water service enters a building through or under the foundation, the pipe must be protected by a sleeve. This sleeve is usually a pipe with a diameter at least twice that of the water service. Once through the foundation, the water service may need to be converted to an acceptable water-distribution pipe. As you learned in reading about approved materials, some materials approved for water-service piping are not approved for interior water distribution.

If the water-service pipe is not made of an acceptable water-distribution material, it must be converted to an approved material, generally within the first 5 feet of its entry into the building. Once inside a building, the hot- and cold-water pipes are referred to as water-distribution pipes. Let's see what you need to know about water-distribution systems.

Trade Tip

A supply pipe must extend to the floor or wall adjacent to the fixture being served. If a reduced-size, flexible water connector is installed between a supply pipe and a fixture, the flexible connector must be an approved product. All flexible water connectors must be accessible.

DISTRIBUTION

Sizing a water-distribution system can become complicated. As we move through this chapter, we will leave sizing exercises until last. There are some rule-of-thumb methods that simplify the sizing of water-distribution pipes. Near the end of the chapter you will get information on sizing a system. For now, we will concentrate on other regulations.

Supplies

Fixture supplies are the tubes or pipes that rise from the fixture branch, the pipe coming out of the wall or floor, to the fixture. A fixture supply may not have a length of more than 30 inches. The required minimum sizing for a fixture supply is determined by the type of fixture supplied with water. A supply pipe must extend to the floor or wall adjacent to the fixture being served. If a reduced-size, flexible water connector is installed between a supply pipe and a fixture, the flexible connector must be an approved product. All flexible water connectors must be accessible.

Pressure-Reducing Valves

Pressure-reducing valves are required to be installed on water systems when the water pressure coming to the water-distribution pipes is in excess of 80 pounds per square inch (psi). The only time this regulation is generally waived is when the water service is bringing water to a device requiring high pressure.

Banging Pipes

Banging pipes are normally the result of water hammer. If you don't want complaining customers, avoid water hammers. You can avoid them in sev-

Trade Tip

Water hammer is most prevalent around quick-closing valves, such as ball cocks and washing-machine valves.

eral ways. You might install air chambers above each faucet or valve. Water-hammer arrestors are available, and do a good job in controlling the problem. These devices are required when a quick-closing valve is used. Expansion tanks can also help with water hammer.

Water hammer is most prevalent around quick-closing valves, such as ball cocks and washing-machine valves. Another way to reduce water hammer is to avoid long, straight runs of pipe. By installing offsets in your water piping, you gradually break up the force of the water. By diminishing the force, you reduce water hammer. All building water-supply systems in which quick-acting valves are installed must be provided with devices to absorb the hammer caused by high pressures resulting from the quick closing of the valves. The devices are to be installed as close as possible to the quick-acting valves.

Booster Pumps

Not all water sources are capable of providing optimum water pressure. When this is the case, a booster pump may be required to increase water pressure. If water pressure fluctuates heavily, the water-distribution system must be designed to operate on the minimum water pressure anticipated.

When calculating the water-pressure needs of a system, you can use information provided by your code book. There swill be ratings for all common fixtures that show the minimum pressure requirements for each type of

Fast Fact

A water-distribution system must be sized to operate satisfactorily under peak demands.

fixture. A water-distribution system must be sized to operate satisfactorily under peak demands.

Booster pumps are required to be equipped with low-water cutoffs. These safety devices are required to prevent the possibility of a vacuum, which may cause backsiphonage.

Water Tanks

When booster pumps are not a desirable solution, water-storage tanks are a possible alternative. Water-storage tanks must be protected from contamination. They may not be located under soil or waste pipes. If the tank is of a gravity type, it must be equipped with overflow provisions.

The water supply to a gravity-style water tank must be automatically controlled. This may be accomplished with a ball cock or other suitable and approved device. The incoming water should enter the tank by way of an air gap. The air gap should be at least 4 inches above the overflow.

Water tanks are also required to have drainage capabilities. The drainpipe must have a valve to prevent draining except when desired.

Pressurized Water Tanks

Pressurized water tanks are the type most commonly encountered in modern plumbing. These tanks are used with well systems. All pressurized water tanks should be equipped with a vacuum-relief valve at the top of the tank. These vacuum-relief valves should be rated for proper operation up to maximum temperatures of 200°F and maximum water pressure of 200 psi. The minimum size of the vacuum-relief valve is 0.5 inch. This requirement is waived for diaphragm/bladder tanks.

Trade Tip

All pressurized water tanks should be equipped with a vacuum-relief valve at the top of the tank.

It is also necessary to equip these tanks with pressure-relief valves. These safety valves must be installed on the supply pipe that feeds the tank or on the tank, itself. The relief valve discharges when pressure builds to a point to endanger the integrity of the tank. The valve's discharge must be carried by gravity to a safe and approved disposal location. The piping carrying the discharge may not be connected directly to the sanitary drainage system.

SUPPORT

The method used to support your pipes is regulated by the plumbing code. There are requirements for the types of materials you may use and how they may be used. Let's see what they are.

One concern with hangers is their compatibility with the pipe they are supporting. You must use a hanger that will not have a detrimental effect on your piping. For example, you may not use galvanized straphanger to support copper pipe. As a rule of thumb, the hangers used to support a pipe should be made from the same material as the pipe being supported. For example, copper pipe should be hung with copper hangers. This eliminates the risk of a corrosive action between two different types of materials. If you are using a plastic or plastic-coated hanger, you may use it with all types of pipe.

The hangers used to support pipe must be capable of supporting the pipe at all times. The hanger must be attached to the pipe and to the member holding the hanger in a satisfactory manner. For example, it would not be acceptable to wrap a piece of wire around a pipe and then wrap the wire around the bridging between two floor joists. Hangers should be securely attached to the member supporting them. For example, a hanger should be attached to the pipe and then nailed to a floor joist. The nails used to hold

CODE UPDATE

Joints between stainless steel and different piping materials must be made with a mechanical joint of the compression or mechanical sealing types or a dielectric fitting.

a hanger in place should be made of the same material as the hanger if corrosive action is a possibility.

Both horizontal and vertical pipes require support. The intervals between supports will vary depending upon the type of pipe being used and whether it is installed vertically or horizontally. The following examples will show you how often you must support the various types of pipes when they are hung horizontally; these examples are the maximum distances allowed between supports in many code jurisdictions:

- ABS: every 4 feet
- Cast iron: every 5 feet
- Galvanized: every 12 feet
- PVC: every 4 feet
- Copper: every 6 feet
- Brass: every 10 feet
- CPVC: every 3 feet
- PB: every 32 feet.

Not all code regions are the same. You may find the following support requirements in place in your region:

- ABS: every 4 feet
- Cast iron: every 15 feet
- Galvanized: every 15 feet
- PVC: every 4 feet
- Copper: every 10 feet
- CPVC: every 3 feet
- Brass: every 10 feet
- PB: every 4 feet.

CODE UPDATE

Hot water is to be piped to the left side of the device dispensing it.

Conservation

Water conservation continues to grow as a major concern. When setting the flow rates for various fixtures, water conservation is a factor. The flow rates of many fixtures must be limited to no more than 3 gallons per minute (gpm). The rating of 3 gpm is based on a water pressure of 80 psi. These fixtures may include the following:

- Showers
- Lavatories
- Kitchen sinks
- Other sinks.

When installed in public facilities, lavatories must be equipped with faucets producing no more than 0.5 gpm. If the lavatory is equipped with a self-closing faucet, it may produce up to 0.25 gpm per use. Water closets are restricted to a use of no more than 4 gallons of water, and urinals must not exceed a usage of 1.5 gallons of water with each use.

Antiscald Precautions

It is easy for the very young or the elderly to receive serious burns from plumbing fixtures. In an attempt to reduce accidental burns, it is required that mixed water to gang showers be controlled by thermostatic means or by pressure-balanced valves. All showers, except for showers in residential dwellings in some regions, must be equipped with pressure-balanced valves or thermostatic controls. These temperature-control valves may not allow water with a temperature of more than 120°F to enter the bathing unit. The maximum water temperature is 110°F in some jurisdictions. Some jurisdictions require safety valves on all showers.

Individual pressure-balancing in-line valves for individual fixture fittings must comply with code requirements. These valves are to be installed in an accessible location and must not be used alone or as a substitute for balanced-pressure, thermostatic, or combination shower valves as required by the code.

Valve Regulations

Gate and ball valves are examples of full-open valves as required under the valve regulations. These valves do not depend on rubber washers, and when they are opened to their maximum capacity, there is a full flow through the pipe. Many locations along the water-distribution installation require full-open valves. These valves may be required in the following locations:

- On the water service before and after the water meter
- On each water service for each building served
- On discharge pipes of water-supply tanks near the tank
- On the supply pipe to every water heater near the heater
- On the main supply pipe to each dwelling.

Some code regions require full-open valves as follows:

- On the water-service pipe near the source connections
- On the main water-distribution pipe near the water service
- On water supplies to water heaters
- On water supplies to pressurized tanks, such as well-system tanks
- On the building side of every water meter.

Full-open valves may be required for use in all water-distribution locations, except as cutoffs for individual fixtures, in the immediate area of the fixtures. There may be other local regulations that apply to specific building uses; check with your local code officer to confirm where full-open valves may be required in your system. All valves must be installed so that they are accessible.

Cutoffs

Cutoff valves do not have to be full-open valves. Stop-and-waste valves are an example of cutoff valves that are not full-open valves. Every sill cock

Fast Fact

Every sill cock must be equipped with an individual cutoff valve.

must be equipped with an individual cutoff valve. Appliances and mechanical equipment that have water supplies are required to have cutoff valves installed in the service piping. Generally, with only a few exceptions, cutoffs are required on all plumbing fixtures. Check with your local code officer for fixtures not requiring cutoff valves. All valves installed must be accessible. Following are some important requirements:

- All devices used to treat or convey potable water must be protected against contamination.
- It is not acceptable to install stop-and-waste valves underground.
- If there are two water systems in a building, one potable, one nonpotable, the piping for each system must be marked clearly. The marking can be in the form of a suspended metal tag or a color code. Your local code may require the pipe to be color-coded and tagged. Nonpotable water piping should not be concealed.
- Hazardous materials, such as chemicals, may not be placed into a potable water system.
- Piping that has been used for a purpose other than conveying potable water may not be used as a potable water pipe.
- Water used for any purpose should not be returned to the potable water supply; this water should be transported to a drainage system.
- Mechanically extracted outlets must have a height not less than three times the thickness of the branch tube wall, and they must be brazed in compliance with the code. Branch tubes must not restrict the flow in the run tube. A dimple/depth stop must be formed in the branch tube to ensure that penetration into the collar is of the correct depth. For inspection purposes, a second dimple must be placed 0.25 inch above the first dimple. The dimples must be aligned with the tube run.
- The minimum required air gap for a fitting is measured vertically from the lowest end of a potable-water outlet to the flood-level rim of the fixture or receptacle in which the potable water outlet discharges.

Fast Fact

Backflow and backsiphonage are genuine health concerns. When a backflow occurs, it can pollute an entire water system.

BACKFLOW

Backflow and backsiphonage are genuine health concerns. When a backflow occurs, it can pollute an entire water system. Without backflow and backsiphonage protection, municipal water services could become contaminated. There are many sources that are capable of deteriorating the quality of potable water. Backflow preventers for hose connections must consist of four independent check valves with an independent atmospheric vent between them and a means of field testing and draining.

Consider this example. A person is using a water hose to spray insecticide on the grounds around a house. The device being used to distribute the insecticide is a bottle-type sprayer attached to a typical garden hose. The bottle has just been filled for use with a poisonous bug killer. A telephone rings inside the home. The individual lays down the bottle sprayer and runs into the house to answer the phone.

While the individual is in the home, leaving the bottle sprayer connected with the sill cock's valve open, a water main breaks. The backpressure caused by the break in the water main creates a vacuum. The vacuum sucks the poisonous contents of the bottle sprayer back into the potable-water system. Now what? How far did the poison go? How much pipe and how many fixtures must be replaced before the water system can be considered safe? The lack of a backflow preventer on the sill cock has created a nightmare. Human health and expensive financial considerations are at stake. A simple, inexpensive backflow preventer could have avoided this potential disaster.

All potable-water systems must be protected against backsiphonage and backflow with approved devices. There are numerous types of devices available to provide this type of protection. The selection of devices will be governed by the local plumbing inspector. It is necessary to choose the proper device for the use intended.

When more than one backflow-prevention valve is installed on a single premise and all are installed in one location, each separate valve must be permanently identified by the permit holder in a manner satisfactory to the administrative authority.

Water-supply inlets to tanks, vats, sumps, swimming pools, and other receptors must be protected by one of the following means: an approved air gap, a listed vacuum breaker installed on the discharge side of the last valve with the critical level not less than 6 inches or in accordance with its listing, or a backflow suitable for the potential contamination or pollution, installed in accordance with the requirements for that type of device or assembly as set forth by the code.

An air gap is the easiest form of protection from backflow. However, air gaps are not always feasible. Since air gaps cannot always be used, there are a number of devices available for the protection of potable-water systems.

Some backflow preventers are equipped with vents. When these devices are used, the vents must not be installed so that they may become submerged. It is also required that these units be capable of performing their function under continuous pressure.

Some backflow preventers are designed to operate in a manner similar to an air gap. With these devices, when conditions occur that may cause a backflow, the devices open and create an open air space between the two pipes connected to it. Reduced-pressure backflow preventers perform this action very well. Another type of backflow preventer that performs on a similar basis is an atmospheric-vent backflow preventer.

Vacuum breakers are frequently installed on water heaters, hose bibbs, and sill cocks. They are also generally installed on the faucet spout of laundry tubs. These devices mount on a pipe or screw onto a hose connection. Some sill cocks are equipped with factory-installed vacuum breakers. These de-

Fast Fact

An air gap is the most positive form of protection from backflow.

Trade Tip

Backflow preventers must be inspected from time to time, so they must be installed in accessible locations.

vices open when necessary and break any siphonic action by introducing air. Hose bibbs must be fitted with nonremovable backflow preventers.

In some specialized cases a barometric loop is used to prevent back-siphonage. These loops must extend at least 35 feet high and can only be used as a vacuum breaker. The loops are effective because they rise higher than the point where a vacuum suction can occur. Barometric loops work on the principle that due to their height suction will not be achieved.

Double-check valves are used in some instances to control backflow. When used in this capacity, double-check valves must be equipped with approved vents. This type of protection would be used on a carbonated-beverage dispenser, for example.

Backflow preventers must be inspected from time to time, so they must be installed in accessible locations.

Some fixtures require an air gap as protection from backflow. Some of these fixtures are: lavatories, sinks, laundry tubs, bathtubs, and drinking fountains. This air gap is accomplished through the design and installation of the faucet or spout serving these fixtures.

When vacuum breakers are installed, they must be installed at least 6 inches above the flood-level rim of the fixture. Vacuum breakers, because of the way they are designed to introduce air into the potable-water piping, may not be installed where they may suck in toxic vapors or fumes. For example, it would not be acceptable to install a vacuum breaker under the exhaust hood of a kitchen range.

When potable water is connected to a boiler for heating purposes, the potable-water inlet should be equipped with a vented backflow preventer. If the boiler water contains chemicals, the potable-water connection should be made with an air gap or a reduced-pressure-principle backflow preventer.

Connections between a potable-water supply and an automatic fire sprinkling system should be made with a check valve. If the potable-water supply is being connected to a nonpotable water source, the connection should be made through a reduced-pressure-principle backflow preventer.

Lawn sprinklers and irrigation systems must be installed with backflow prevention in mind. Vacuum breakers are a preferred method for backflow prevention, but other types of backflow preventers are allowed.

Cross-connections are prohibited, except where approved protective devices are installed. Water pumps, filters, softeners, tanks, and all other devices that handle or treat potable water must be protected from contamination. Pressure-type vacuum breakers must conform to ASSE 1020 for outdoor use.

All pullout spout-type faucets must be in compliance with CSA CAN/CSA-B125 and have an integral vacuum breaker or vent to atmosphere or a dedicated deck- or wall-mounted vacuum breaker in their design. All faucets with integral atmospheric or spillproof vacuum breakers must be installed in accordance with the manufacturer's recommendations.

Water heaters and boiler drain valves that are provided with hose-connection threads used only for draining the water heater do not require backflow protection. The water-connection devices for automatic clothes washers do not require backflow protection when such protection is provided inside the washing machine.

The potable-water supply for a carbonator must be protected by either an air gap or a vented backflow preventer. Carbonated-beverage dispensers without an approved internal air gap or appropriate vented backflow preventer must protect the water supply itself with a vented backflow preven-

Trade Tip

Water heaters and boiler drain valves that are provided with hose-connection threads used only for draining the water heater do not require backflow protection.

Fast Fact

When hot-water pipe is installed, it is expected to maintain the temperature of its hot water for a distance of up to 100 feet from the fixture it serves.

ter. The backflow-preventer device and the piping downstream must not be affected by carbon-dioxide gas.

Potable-water supplies for fire-protection systems that are normally under pressure, except in one- and two-family residential sprinkler systems with piped-in materials approved for potable-water distribution, must be protected from backpressure and backsiphonage. This protection can be in the form of a double check-valve assembly, a double check-detector assembly, a reduced-pressure backflow preventor, or a reduced-pressure detector assembly. Potable-water supplies to fire-protection systems that are not normally under pressure must be protected from backflow and must meet the requirements of the appropriate standards.

Chemical dispensers require backflow devices that comply with ASSE 1055 and must be equipped with an air-gap fitting.

Systems that are under continuous pressure and contain chemical additives or antifreeze or systems connected to a nonpotable secondary water supply must be equipped with approved devices to protect potable-water systems. A reduced-pressure-principle backflow preventer is required for this type of situation. Sometimes chemical additives or antifreeze is added to only a portion of an automatic fire-sprinkler or standpipe system. When this is the case, the backflow preventer can be located so as to isolate only that portion of the system. Systems that are not under continuous pressure can be protected with an air gap or a pipe-applied atmospheric vacuum breaker.

HOT-WATER INSTALLATIONS

When hot-water pipe is installed, it is expected to maintain the temperature of its hot water for a distance of up to 100 feet from the fixture it serves. If the distance between the hot-water source and the fixture being served is more than 100 feet, a recirculating system is frequently required. When a recirculating system is not appropriate, other means may be used to maintain water temperature. These means could include insulation or heating tapes. Check with your local code officer for approved alternates to a recirculating system if necessary.

If a circulator pump is used on a recirculating line, the pump must be equipped with a cutoff switch. The switch may operate manually or automatically. Piping for this type of system is required to be insulated in accordance with the International Energy Conservation Code.

Hot water to fixtures must be installed in such a manner that the control for the hot water will be to the left-hand side of the fixture fitting. As an exception to this rule, a mixing valve for a shower or tub-shower combination can rely on distinctive markings on the device to indicate the proper position for the distribution of hot water.

Residential occupancies require hot water for all fixtures and equipment used for bathing, washing, cooking, cleaning, laundry, and building maintenance. Nonresidential occupancies require hot water for culinary purposes, cleaning, laundry, and building maintenance. Hot water must be tempered in nonresidential uses for bathing and washing and must be delivered from accessible hand-washing facilities.

WATER HEATERS

The standard working pressure for a water heater is 125 psi. The maximum working pressure of a water heater is required to be permanently marked in an accessible location. Every water heater is required to have a drain, located at the lowest possible point on the water heater. Some exceptions may be allowed for very small, under-the-counter water heaters.

Fast Fact

Relief valves are mandatory equipment on water heaters.

All water heaters are required to be insulated. The insulation factors are determined by the heat loss of the tank in an hour's time. These regulations are required of a water heater before it is approved for installation.

Relief valves are mandatory equipment on water heaters. These safety valves are designed to protect against excessive temperature and pressure. The most common type of safety valve used will protect against both temperature and pressure from a single valve. The blow-off rating for these valves must not exceed 210°F and 150 psi. The blow-off rating for the pressure-relief valve must not exceed the maximum working pressure of the water heater it serves, usually 125 psi.

When temperature- and pressure-relief valves are installed, their sensors should monitor the top 6 inches of water in the water heater. There may not be any valves located between the water heater and the temperature- and pressure-relief valves.

The blow-off from relief valves must be piped down to protect bystanders in the event of a blow-off. The pipe used for this purpose must be rigid and capable of sustaining temperatures of up to 210°F. The discharge pipe must be the same size as the relief valve's discharge opening, and it must run, undiminished in size, to within 6 inches of the floor. If a relief-valve discharge pipe is piped into the sanitary drainage system, the connection must be through an indirect waste. The end of a discharge pipe may not be threaded, and no valves may be installed in the discharge pipe.

When the discharge from a relief valve may damage property or people, safety pans should be installed. These pans typically have a minimum depth of 1.5 inches. Plastic pans are commonly used for electric water heaters, and metal pans are used for fuel-burning heaters. These pans must be large enough to accommodate the discharge flow from the relief valve.

The pan's drain may be piped to the outside of the building or to an indirect waste where people and property will not be affected. The discharge location should be chosen so that it will be obvious to building occupants when a relief valve discharges. Traps should not be installed on the discharge piping from safety pans.

Water heaters must be equipped with an adjustable temperature control. This control is required to be automatically adjustable from the lowest to the highest temperatures allowed. Some locations restrict the maximum water temperature in residences to 110 or 120°F. There must be a switch to shut off the power to electric water heaters. When the water heater uses a fuel, such as gas, there must be a valve available to cut the fuel source off. Both the electric and fuel shutoffs must be able to operate without affecting the remainder of the building's power or fuel. All water heaters requiring venting must be vented in compliance with local code requirements.

PURGING

When a potable-water system has been installed, added to, or repaired, it must be disinfected. For years this amounted to little more than running water through the system until it appeared clean. This is no longer enough. Under today's requirements, the system must undergo a true cleansing procedure.

The precise requirements for clearing a system of contaminants will be prescribed by the local health department or code-enforcement office. Typically, it will require flushing the system with potable water until the water appears clean. This action will be followed by cleaning with a chlorine solution. The exact requirements for the mixture of chlorine and water will be provided by a local agency. The chlorine mixture is introduced into the system and normally remains between 3 and 24 hours.

After the chlorine has been in the system for the required time, the system is flushed with potable water until there is no trace of chlorine remaining. Again, check with your local authorities on the exact specifications for purifying the potable-water system.

Trade Tip

The quality of water from a private source must meet minimum standards as potable water.

CODE UPDATE

Backflow preventers shall not be located in areas subject to freezing, unless the device can be installed with unions for removal.

WELLS

When you will be working with wells or other private water sources, there are some rules you must follow. This section will apprise you of what you need to know when working with private water systems.

If a building does not have access to a public water source, it must depend on water from a private source. Typically, this source is a well, but in some cases it could be a cistern, spring, or stream. If surface water is used as a potable-water source, it must be tested and approved for use. For that matter, wells are generally required to be tested and approved.

The quality of water from a private source must meet minimum standards as potable water. This is determined through water tests. The determination of potability is normally done by the local health department or some other local authority.

The quantity of water delivered from a well must also meet certain requirements. The well or water source must be capable of supplying enough water for the intended use of the system. For example, a single-family home may be rated as requiring 75 gallons of water a day for each occupant. Hospitals, on the other hand, are rated to require a minimum of between 150 and 250 gallons of water each day for each bed in the hospital.

All private water sources must be protected from contamination. They must also be disinfected before being used. The protection from contami-

Trade Tip

All wells should be located above and upstream from any possible contaminating sources, such as a septic field.

nation can include several features. For example, wells must have watertight caps. They may not be installed in an area where contamination is likely, such as near a septic system. The height of a well casing should extend above the ground. All wells should be located above and upstream from any possible contaminating sources.

Any well, whether drilled, dug, or driven, generally must not deliver water for potable use from a depth of less than 10 feet. There are rules governing the allowable distances between a private water source and possible pollution sources. The following examples show required spacing from a few of the possible polluting sources:

- Septic tank: 25 feet
- Sewer: 10 feet
- Pasture: 100 feet
- Barnyard: 100 feet
- Underground disposal fields: 50 feet.

Well Requirements

Wells must be installed to meet minimum standards. What follows is a description of the minimum requirements for the installation of various types of wells in a common code area.

Dug and Bored Wells

Dug and bored wells are usually relatively shallow. Their casings are required to be made from waterproof concrete, corrugated metal pipe, galvanized pipe, or tile. These casings must extend to a minimum depth of 10 feet below the ground. The casing is required to extend below the water table. For example, if the well is 16 feet deep and the water table is 13 feet deep, the casing must extend at least 13 feet into the ground.

When wells are dug or bored, there is a large space between the well casing and the undisturbed earth. This space is required to be filled with a grout material. The grout must encompass the well casing and have a minimum thickness of 6 inches. This helps to prevent surface water from entering the well. It is also necessary for the well casing to rise at least 6 inches above the well platform.

This type of well is usually large in diameter. The top of the well must be sealed with a watertight cover. Covers must overlap the sides of the well casing and extend downwards for a minimum of 2 inches. Concrete covers are common on this type of well.

Common practice with bored and dug wells is to have the water pipe from the well to the house exit through the side of the casing. This is generally done below ground and below the regional frost line. Where penetration of the well casing occurs, the hole must be sealed to prevent outside water from entering the well.

If a well is in an area subject to flooding, the well casing and cover must be designed and installed to withstand the forces associated with a flood. Grading of the ground surrounding a private water source may be required to divert runoff water from entering the potable-water supply.

Drilled and Driven Wells

Drilled and driven wells are different from dug and bored wells. The diameters of these wells are much smaller, and drilled wells are often very deep. The casings for these wells must be made of steel or some other approved material. The casing must extend at least 6 inches above the well platform.

Grouting is required around the exterior of these casings. The grout material is required to extend a minimum depth of 10 feet or solid contact with rock, whichever comes first. The casing should extend into rock or well beyond the water-table level.

Extending the water pipe from the well to a building is usually done through the side of the casing. Normal procedure calls for the use of a pitless adapter in these installations. Pitless adapters mount into the well casing and form a watertight seal. In any event, the casing must be sealed at any openings that might allow nonpotable water to enter the well.

The cover for a drilled well must be waterproof and will usually allow for electrical wires connected to the pump. These wires must get from the submerged pump to the control box. In all cases, the cover must be designed and installed to prevent the influx of surface water into the well.

Well Pumps

The pumps used with potable-water systems must meet minimum standards. This section identifies and explains these standards. Pumps must be approved for use. They must be readily accessible for service, maintenance, or repair. In flood-prone areas, pumps must be designed and installed to resist the potential detrimental effects of a flood. Water pumps are required to be capable of continuous operation.

In some jurisdictions, if a pump is installed in a home, it must be installed on an appropriate base. Some pumps are installed on brackets, connected to pressure-supply tanks. If a pump is installed in a basement, it must be installed at least 18 inches above the basement floor. It is not acceptable to install the pump in a pit or closer than 18 inches to the finished floor level of a basement floor. This provision is meant to protect the pump from submersion through basement flooding.

Pump Houses

When shallow-well pumps or two-pipe jet pumps are used, the pumps are sometimes placed in pump houses. These pump houses must be of approved construction. A building providing shelter to a water pump must be equipped to prevent the pump or related piping from freezing. Such an enclosure must also be provided with adequate drainage facilities to prevent water from rising over the pump and piping.

SIZING

About the only aspect of the potable-water system that we have not covered is sizing. This section of the chapter will show you how to size potable-water piping. But, be advised: this procedure is not always simple, and it requires concentration. If your mind is not fresh, leave this section for a later reading. If you are ready to learn how to size water pipe, get a pen and paper and get ready for one of the more complicated aspects of this book.

Some facets of potable-water pipe sizing are not very difficult. Many times your code book will provide charts and tables to help you. Some of these graphics will detail precisely which size pipe or tubing is required. But, unfortunately, code books cannot provide concrete answers for all piping installations.

Many factors affect the sizing of potable-water piping. The type of pipe used will have an influence on your findings. Some pipe materials have smaller inside diameters than others. Some pipe materials have a rougher surface or more restrictive fittings than others. Both of these factors will affect the sizing of a water system.

When sizing a potable-water system, you must be concerned with the speed of the flowing water, the quantity of water needed, and the restrictive qualities of the pipe being used to convey the water. Most materials approved for potable-water piping will provide a flow velocity of 5 feet per second. The exception is galvanized pipe, which provides a speed of 8 feet per second.

It may be surprising that galvanized pipe allows a faster flow rate. This occurs because of the wall strength of galvanized pipe. In softer pipes, such as copper, fast-moving water can essentially wear a hole in the pipe. These flow ratings are not carved in stone. I am sure you will find people who will argue for either a higher or a lower rating, but these ratings are in use with current plumbing codes.

When considering the three factors previously discussed to determine pipe size, you must use math that you may not have seen since your school days, and you may not have seen it then. Let me give you an example of how a typical formula might look. A common formula could resemble the one listed below:

X = the water's rate of flow—in most cases 5 feet per second

Y = the quantity of water in the pipe

Z = the inside diameter of the pipe

A typical formula might look like this:

$$Y = XZ$$

Since most plumbers will refuse to do this type of math, most code books offer alternatives. The alternatives are often in the form of tables or charts that show pertinent information on the requirements for pipe sizing.

The tables or charts you might find in a code book are likely to discuss the following: a pipe's outside diameter, a pipe's inside diameter, a flow rate for the pipe, and a pressure loss in the pipe over a distance of 100 feet. These charts or tables will be dedicated to a particular type of pipe. For example, there would be one table for copper pipe and another table for PEX pipe.

The information supplied in a ratings table for PEX pipe might look like this:

- Pipe size is 0.75 inch
- Inside pipe diameter is 0.715
- The flow rate, at 5 feet per second, is 6.26 gpm
- Pressure lost in 100 feet of pipe is 14.98.

This type of pipe sizing is most often done by engineers, not plumbers. When sizing a potable-water system, the sizing exercise starts at the last fixture and works its way back to the water service.

Commercial jobs, in which pipe sizing can become quite complicated, are generally sized by design experts. All a working plumber is required to do is install the proper pipe sizes in the proper locations and manner. For residential plumbing, where engineers are less likely to have a hand in the design, there is a rule-of-thumb method to sizing most jobs. In the average home, a 0.75-inch pipe is sufficient for the main artery of the water-distribution system. Normally, not more than two fixtures can be served by a 0.5-inch pipe. With this in mind, sizing becomes simple.

The 0.75-inch pipe is normally run to the water heater, and it is typically used as a main water-distribution pipe. When nearing the end of a run, the 0.75-inch pipe is reduced to 0.5-inch pipe where there are no more than two fixtures to connect to. Most water services will have a 0.75-inch diam-

CODE UPDATE

The termination of piping from a relief port or air gap fitting for a backflow preventer must discharge into an approved indirect waste receptor or to the outdoors where it will not cause damage or create a nuisance.

eter, with those serving homes with numerous fixtures using a 1-inch pipe. This rule-of-thumb sizing will work on almost any single-family residence.

The water supplies to fixtures are required to meet minimum standards. These sizes are derived from local code requirements. You simply find the fixture you are sizing and check the column heading for the proper supply size.

Most code requirements seem to agree that there is no definitive way to set a boilerplate formula for establishing potable-water pipe sizing. Code officers can require pipe sizing to be performed by a licensed engineer. In most major plumbing systems the pipe sizing is done by a design professional.

Code books give examples of how a system might be sized, but the examples are not meant as a code requirement. The code requires a water system to be sized properly. However, due to the complexity of the process, the books do not set firm, precise requirements for the process. Instead, they give parts of the puzzle in the form of some minimum standards, but it is up to a professional designer to come up with an approved system.

Where does this leave you? The sizing of a potable-water system is one of the most complicated aspects of plumbing. Very few single-family homes are equipped with potable-water systems designed by engineers. I have already given you a basic rule-of-thumb method for sizing small systems. Next, I am going to show you how to use the fixture-unit method of sizing.

The fixture-unit method is not very difficult, and it is generally acceptable to code officers. While this method may not be perfect, it is much faster and easier to use than the velocity method. Other than additional expense in materials, you can't go wrong by oversizing your pipe. If in doubt on sizing, go to the next larger size. Let's see how you might size a single-family residence's potable-water system using the fixture-unit method.

CODE UPDATE

Water supply connections to beverage dispensers must be protected against backflow.

Most codes assign a fixture-unit value to common plumbing fixtures. To size with the fixture-unit method, you must establish the number of fixture units to be carried by the pipe. You must also know the working pressure of the water system. Most codes will provide guidelines for these two pieces of information.

For our example, we have a house with the following fixtures: three toilets, three lavatories, one bathtub-shower combination, one shower, one dishwasher, one kitchen sink, one laundry hookup, and two sill cocks. The water pressure serving this house is 50 psi. There is a 1-inch water meter serving the house, and the water service is 60 feet in length. With this information and the guidelines provided by your local code, you can do a pretty fair job of sizing your potable-water system.

The first step is to establish the total number of fixture units on the system. The code regulations will provide this information. You have three toilets—nine fixture units. The three lavatories add three fixture units. The tub-shower combination counts as two fixture units; the showerhead over the bathtub doesn't count as an additional fixture. The shower adds two fixture units. The dishwasher adds two fixture units, and so does the kitchen sink. The laundry hookup counts as two fixture units. Each sill cock is worth three fixture units. This house has a total fixture-unit load of 28. Now you have the first piece of your sizing puzzle solved. The next step is to determine what size pipe will allow your number of fixture units.

Our subject house has a water pressure of 50 psi. This pressure rating falls into the category allowed by code. First, find the proper water-meter size; the one you are looking for is 1 inch. You will notice that a 1-inch meter and a 1-inch water service are capable of handling sixty fixture units when the pipe is only running 40 feet. However, when the pipe length is stretched to 80 feet, the fixture load is dropped to forty-one. At 200 feet, the fixture rating is twenty-five. What is it at 100 feet? At 100 feet, the allowable fixture load is thirty-six. See, this type of sizing is not so hard.

What does this tell us? We know the water service is 60 feet long. Once inside the house, how far is it to the most remote fixture? In this case, the furthest fixture is 40 feet from the water-service entrance. This gives us a developed length of 100 feet, 60 feet for the water service and 40 feet for the interior pipe. Going to the table in our code book, we see that for 100 feet of pipe, under the conditions in this example, we are allowed 36 fixture units. The house only has 28 fixture units, so our pipe sizing is fine.

What would happen if the water meter were a 0.75-inch meter instead of a 1-inch meter? With a 0.75-inch meter and a 1-inch water service and main distribution pipe, we could have thirty-three fixture units. This would still be a suitable arrangement, since we only have 28 fixture units. Could we use a 0.75-inch water service and water-distribution pipe with the 0.75-inch meter? No, we couldn't. With all sizes set at 0.75 inch, the maximum number of fixture units allowed is 17.

In this example, the piping is oversized. But if you want to be safe, use this type of procedure. If you are required to provide a riser diagram showing the minimum pipe sizing allowed, you will have to do a little more work. Once inside a building, water-distribution pipes normally extend for some distance, supplying many fixtures with water. As the distribution pipe continues on its journey, it reduces the fixture load as it goes.

For example, assume that the distribution pipe serves a full bathroom group within 10 feet of the water service. Once this group is served with water, the fixture-unit load on the remainder of the water-distribution piping is reduced by six fixture units. As the pipe serves other fixtures, the fixture-unit load continues to decrease. So it is feasible for the water-distribution pipe to become smaller as it goes along.

Let's take our same sample house and see how we could use smaller pipe. Okay, we know we need a 1-inch water service. Once inside the foundation, the water service becomes the water-distribution pipe. The water heater is located 5 feet from the cold-water distribution pipe. The 1-inch pipe will extend over the water heater and supply it with cold water. There will be a hot-water distribution pipe originating at the water heater. Now you have two water-distribution pipes to size.

When sizing the hot and cold water pipes, you could make adjustments for fixture-unit values on some fixtures. For example, a bathtub is rated as two fixture units. Since the bathtub rating is inclusive of both hot and cold water, obviously the demand for just the cold-water pipe is less than that shown in our table. For simplicity, I will not break the fixture units down into fractions or reduced amounts. I will show you the example as if a bath-

CODE UPDATE

Potable water outlets and stop-and-waste valves shall not be installed below ground or grade.

tub required two fixture units of hot water and two fixture units of cold water. However, you could reduce the amounts listed in the table by about 25 percent to obtain the rating for each hot- and cold-water pipe. For example, the bathtub, when being sized for only cold water, could take on a value of 1.5 fixture units.

Let's get on with the exercise. We are at the water heater. We ran a 1-inch cold-water pipe overhead and dropped a 0.75-inch pipe into the water heater. What size pipe do we bring up for the hot water? First, count the number of fixtures that use hot water, and assign them a fixture-unit value. The fixtures using hot water are all fixtures, except the toilets and sill cocks. The total count for hot-water fixture units is lucky number thirteen. From the water heater, our most remote hot-water fixture is 33 feet away.

What size pipe should we bring up from the water heater? By looking at the table in our code book, we find a distance and fixture-unit count that will work in this case. You would look under the 40-feet column, since our distance is less than 40 feet. When you look in the column, the first fixture-unit number you see is nine; this won't work. The next number is twenty-seven; this one will work, because it is greater than the thirteen fixture units we need. Looking across the table, you will see that the minimum pipe size to start with is a 0.75-inch pipe. Isn't it convenient that the water heater just happens to be sized for 0.75-inch pipe?

Now we start our hot-water run with 0.75-inch pipe. As our hot-water pipe goes along the 33-foot stretch, it provides water to various fixtures. When the total fixture count remaining to be served drops to less than nine, we can reduce the pipe to 0.5-inch pipe. We can also run our fixture branches off the main using 0.5-inch pipe. We can do this because the highest fixture-unit rating on any of our hot-water fixtures is two fixture units. Even with a pipe run of 200 feet we can use 0.5-inch pipe for up to four fixture units. Is this sizing starting to ring a bell? Remember the rule-of-thumb siz-

CODE UPDATE

Buildings that contain piping for nonpotable water must have the pipe labeled as nonpotable water with either color markings or metal tags. The color purple is used to identify pipes that are conveying reclaimed water, rain, or gray water.

ing I gave you earlier? These sizing examples are making the rule-of-thumb method ring true.

With the hot-water sizing done, let's look at the remainder of the cold water. We have less than 40 feet to our furthest cold-water fixture. We branch off near the water-heater drop for a sill cock, and there is a full bathroom group within 7 feet of our water-heater drop. The sill-cock branch can be 0.5-inch pipe. The pipe going under the full bathroom group could probably be reduced to 0.75 pipe, but it would be best to run it as a 1-inch pipe. However, after serving the bathroom group and the sill cock, how many fixture units are left? There are only nineteen fixture units left. We can now reduce to 0.75-inch pipe, and when the demand drops to below nine fixture units we can reduce to 0.5-inch pipe. All of our fixture branches can be run with 0.5-inch pipe.

This is one way to size a potable-water system that works without driving you crazy. There may be an argument against the sizes I gave in these examples. The argument would be that some of the pipe is oversized, but, as I said earlier, when in doubt, go bigger. In reality, the cold-water pipe in the last example could probably have been reduced to 0.75-inch pipe where the transition was made from water-service to water-distribution pipe. It could almost certainly have been reduced to 0.75-inch pipe after the water-heater drop. Local codes will have their own interpretation of pipe sizing, but this method will normally serve you well. Always refer to your local code book for specific sizing requirements and practices.

FACTS TO KEEP YOU OUT OF TROUBLE

Here are some more facts to keep you out of trouble when you are working with potable-water systems:

- When working with a solvent-cement joint, you are not required to use a primer when all of the following conditions apply: the cement being used is third-party certified as conforming to ASTM F493, the cement being used is yellow in color, the cement is used only for joining 0.5-inch to 2-inch diameter CPVC pipe and fittings, and the CPVC pipe and fittings are manufactured in accordance with ASTM D2846.
- Cross-linked polyethylene plastic requiring joints between tubing or fittings must comply with the plumbing code.
- PEX tubing must be marked appropriately to identify the uses that the material is approved for.
- PEX fittings must be made of metal and must be secured with metal compression fittings.
- PEX tubing may not be used for water-heater connections within the first 18 inches of the piping connected to the water heater.
- Flared pipe ends must be made with a tool designed for that operation.
- Mechanical joints must be made in accordance with the local plumbing code and the manufacturer's recommendations. Metallic lock rings and insert fittings as described in ASTM F1807 are required for the installation.
- Tempered water is required to be delivered from accessible hand-washing facilities.
- Methods for maintaining energy efficiency must conform to the International Energy Conservation Code.
- Vacuum breakers for hose connections in healthcare or laboratory areas must not be less than 6 feet above the floor.
- Pipe identification must include the contents of the piping system and an arrow indicating the direction of flow. Any hazardous piping systems must contain information that addresses the nature of the hazard. Pipe identification must be repeated at maximum intervals of 25 feet and at each point where the piping passes through a wall, floor, or roof. All lettering must be readily observed within the room or space

where the piping is located. Any coloring for pipe identification must be discernible and consistent throughout the building. Identification labeling must be sized in compliance with code requirements.

- Pressure-type vacuum breakers must not be installed in locations where spillage could cause damage to the structure.
- The discharge from a reverse-osmosis system must enter a drainage system through an air gap or an air-gap device.
- Construction, installation, alterations, and repair of solar systems, equipment, and appliances intended to utilize solar energy for space heating or cooling, domestic hot-water heating, swimming-pool heating, or process heating must be in accordance with the International Mechanical Code.
- Flexible corrugated connectors made of copper or stainless steel are limited in their length. Water-heater connectors must not be more than 24 inches long. Fixture connectors are limited to a maximum length of 30 inches. Connectors for washing machines must not exceed 72 inches in length. Flex connectors for dishwashers and icemakers must not be longer than 120 inches.
- Female PVC screwed fittings for water piping must be used with plastic male fittings and plastic male threads only.
- Joints between copper tubing and galvanized-steel pipe must be made with a brass or dielectric fitting. The copper tubing must be soldered to the fitting in an approved manner. Galvanized-steel pipe is to be screwed into the connector.

This concludes our view of water distribution. It is now time to cover the subject of drainage systems. We will do this in the next chapter.

CHAPTER 7

Sanitary Drainage Systems

Drainage systems intimidate many people. When they look at their code books, they see charts and math requirements that make them nervous. Their fear is largely unjustified. For the inexperienced, the fundamentals of building a suitable drainage system can appear formidable. But with a basic understanding of plumbing, the process becomes much less complicated. This chapter is going to take you step-by-step through the procedures for making a working drainage system.

You are going to learn the criteria for sizing pipe. You will be shown which types of fittings can be used in various applications. During the process, you will be given instructions for the proper installation of a drainage system.

SIZING

Sizing pipe for a drainage system is not difficult. To size pipe for drainage, there are a few benchmark numbers you must know, but you don't have to memorize them. Your code book will have charts and tables that provide the benchmarks. All you must know is how to interpret and use the information provided.

The size of a drainage pipe is determined by using various factors, the first of which is the drainage load. This refers to the volume of drainage the pipe will be responsible for carrying. When you refer to your code book, you will find ratings that assign a fixture-unit value to various plumbing fixtures. For example, a residential toilet has a fixture-unit value of four. A bathtub's fixture-unit value is two.

By using the ratings given in your codebook, you can quickly assess the drainage load for the system you are designing. Since plumbing fixtures require traps, you must also determine which size traps are required for particular fixtures. Again, you don't need a math degree to accomplish this task. In fact, your code book will tell you which trap sizes are required for most common plumbing fixtures.

Your code book will provide trap-size requirements for specific fixtures. For example, by referring to the ratings in your book, you will find that a bathtub requires a 1.5-inch trap. A lavatory needs a 1.25-inch trap. The list describes the trap needs for all common plumbing fixtures. Trap sizes will not be provided for toilets, since toilets have integral traps.

When necessary, you can determine a fixture's drainage-unit value by the size of the fixture's trap. A 1.25-inch trap, the smallest trap allowed, will carry a fixture-unit rating of one. A 1.5-inch trap will have a fixture unit of two. A 2-inch trap will have a rating of three fixture units. A 3-inch trap will have a fixture-unit rating of five, and a 4-inch trap will have a fixture-unit rating of six. This information can be found in your code book and may be applied to a fixture not specifically listed with a rating in the book.

Determining the fixture-unit value of a pump does require a little math, but it's simple. Start by taking the flow rate in gallons per minute (gpm), and as-

Trade Tip

When necessary, you can determine a fixture's drainage-unit value by the size of the fixture's trap.

Trade Tip

A normal grade is generally 0.25 inch to the foot, but the fall can be steeper or shallower.

sign two fixture units for every gpm of flow. For example, a pump with a flow rate of 30 gpm would have a fixture-unit rating of sixty. Some code jurisdictions are more generous. For example, you may find that your local code will allow one fixture unit to be assigned for every 7.5 gpm. With the same pump, producing 30 gpm, the liberal fixture-unit rating would be four. That's quite a difference from the ratings in more conservative code regions.

Other considerations when sizing drainage pipe are the type of drain you are sizing and the amount of fall that the pipe will have. For example, sizing a sewer is a different from sizing a vertical stack. A pipe with a .25-inch fall is rated differently than the same pipe with a .125-inch fall.

Building Drains

Building drains and sewers use the same criteria in determining the proper pipe size. The two components you must know to size these types of pipes are the total number of drainage fixture units entering the pipe and the amount of fall placed on the pipe. The amount of fall is based on how much the pipe drops in each foot it travels. A normal grade is generally 0.25 inch to the foot, but the fall can be steeper or shallower. Drainage fixture-unit values for continuous and semi-continuous flow into a drainage system are computed on the basis that 1 gpm of flow is equal to two fixture units. Keep in mind that provisions for any future fixtures must be taken into account when sizing a plumbing system.

When you refer to your code book, you will find information, probably a table, to aid you in sizing building drains and sewers. Let's take a look at how a building drain for a typical house would be sized.

Sizing Example

Our sample house has two and one-half bathrooms, a kitchen, and a laundry room. To size the building drain for this house, we must determine the total fixture-unit load that may be placed on the building drain. To do this, we start by listing all the plumbing fixtures producing a drainage load. In this house we have the following fixtures:

- One bathtub
- One shower
- Three toilets
- Three lavatories
- One kitchen sink
- One dishwasher
- One clothes washer
- One laundry tub.

By using the chart in my local code book, I can determine the number of drainage fixture units assigned to each of these fixtures. When I add up all the fixture units, the total load of 28 is established. It is always best to allow a little extra in your fixture-unit load so your pipe will be in no danger of becoming overloaded. The next step is to look at the chart in the code book to determine the sizing of the building drain.

The building drain will be installed with a 0.25-inch fall. By looking at the chart in the code book, we see that we can use a 3-inch pipe for our building drain, based on the number of fixture units, but a notice in the footnote below the chart indicates that a 3-inch pipe may not carry the discharge of more than two toilets and out test house has three toilets. This means that we will have to move up to a 4-inch pipe.

Suppose our test house had only two toilets. If we eliminate one of the toilets, our fixture load drops to 24. According to the table, we could use a 2.5-inch pipe, but we know our building drain must use at least a 3-inch pipe to connect to the toilets. A fixture's drain may enter a pipe the same size as

Fast Fact

Horizontal branches are the pipes branching off from a stack to accept the discharge from fixture drains.

the fixture drain or a pipe that is larger, but it may never be reduced to a smaller size except with a 4-by-3-inch closet bend.

So, with two toilets, our sample house could have a building drain and sewer with a 3-inch diameter. But should we run a 3-inch pipe or a 4-inch pipe? In a highly competitive bidding situation, 3-inch pipe would probably win the coin toss. It would be less expensive to install a 3-inch drain, and you would be more likely to win the bid on the job. However, when feasible, it would be better to use a 4-inch drain. This allows the homeowner to add another toilet at some time in the future. If you install a 3-inch sewer, the homeowner would not be able to add a toilet without replacing the sewer with 4-inch pipe.

Horizontal Branches

Horizontal branches are the pipes branching off from a stack to accept the discharge from fixture drains. Horizontal branches normally leave the stack as a horizontal pipe, but they may turn to a vertical position and still retain the term "horizontal branch."

The procedure for sizing a horizontal branch is similar to that used to size a building drain or sewer, but the ratings are different. Your code book will contain the benchmarks for your sizing efforts, but let me give you some examples.

The number of fixture units allowed on a horizontal branch is determined by pipe size and pitch. All the following examples are based on a pitch of 0.25 inch to the foot. A 2-inch pipe can accommodate up to six fixture units in most code regions. Some jurisdictions allow up to eight fixture units. A 3-inch pipe can handle 20 fixture units but not more than two toilets. A 3-inch pipe is allowed up to 35 units and up to three toilets in some jurisdictions. A 1.5-inch pipe will carry either two or three fixture units, depending upon your local code requirements. When the additional fixture unit is allowed, it may not be from sinks, dishwashers, or urinals. A 4-inch pipe will take up to 160 fixture units in most code areas. You may find a jurisdiction that will allow up to 216 units.

Stack Sizing

Stack sizing is not too different from the other sizing exercises we have studied. When you size a stack, you must base your decision on the total

Fast Fact

When you size a stack, you must base your decision on the total number of fixture units carried by the stack and the amount of discharge into branch intervals.

number of fixture units carried by the stack and the amount of discharge into branch intervals. This may sound complicated, but it isn't.

There are tables in your local code book that help you with sizing pipes. You will notice that there are three columns. The first is for pipe size, the second represents the discharge of a branch interval, and the last column shows the ratings for the total fixture-unit load on a stack. This table is based on a stack with no more than three branch intervals.

Sizing the stack requires you to first determine the fixture load entering the stack at each branch interval. Let me give you an example of how this type of sizing works. In our example we will size a stack that has two branch intervals. The lower branch has a half-bath and a kitchen on it. Using the ratings from common code regions, the total fixture-unit count for this branch is six. This is determined from the table providing ratings for various fixtures.

The second stack has a full bathroom group on it. The total fixture-unit count on this branch is six, if you use a bathroom-group rating, or seven, if you count each fixture individually. I would use the larger of the two numbers.

When you look at the table in your code book, you will see the horizontal listings for a 3-inch pipe. You know the stack must have a minimum size of 3 inches to accommodate the toilets. As you look across the table, you will see that each 3-inch branch may carry up to twenty fixture units. Your first branch has six fixture units and the second branch has seven fixture units, so both branches are within their limits.

When you combine the total fixture units from both branches, you have a total of thirteen. Continuing to look across the table, you see that the stack can accommodate up to 48 fixture units. Obviously, a 3-inch stack is adequate for your needs. If the fixture-unit loads had exceeded the numbers in either of the columns, the pipe size would need to be increased.

When sizing a stack, it is possible that the developed length of the stack will be comprised of different sizes of pipe. For example, at the top of the stack the pipe size may be 3 inches, but at the bottom the pipe size may be 4 inches. This is because as you get to the lower portion of the stack, the total fixture-unit load placed on the stack is greater. Remember to check your local code requirements, since they may be different from the ones I am working with.

INSTALLATIONS

Once the pipe is properly sized, it is ready for installation. There are a few regulations pertaining to pipe installation that you need to be aware of.

Grading

When you install horizontal drainage piping, it must fall toward the waste-disposal site. A typical grade for drainage pipe is 0.25 inch of fall per foot. This means that the lower end of a 20-foot piece of pipe would be 5 inches lower than the upper end when properly installed. While the 0.25-inch-to-the-foot grade is typical, it is not the only acceptable grade for all pipes.

If you are working with pipe that has a diameter of 2.5 inches, or less, the minimum grade for the pipe is 0.25 inch to the foot. Pipes with diameters between 3 and 6 inches are allowed a minimum grade of 0.125 inch to the foot. Some code zones require special permission to be granted prior to installing pipe with a 0.125-inch-to-the-foot grade. Pipes with diameters of 8 inches or more may be allowed to be installed with an acceptable grade of 0.0625 inch to the foot.

Trade Tip

Schedule 80 or heavier pipe can be threaded with dies specifically designed for plastic piping.

JOINTS

There are a number of requirements for making joints between different types of pipes and fittings. Of course, all connections must be made in compliance with the local plumbing code. There are some aspects of joining pipes that are very specific, and we are going to talk about them here.

Mechanical joints on drainage pipes must be made with elastomeric seal. They may only be used on underground piping and must comply with the manufacturer's recommendations.

Solvent-weld joints are required to be made with pipe whose surface ends are clean and free from moisture. Joints must be made while the cement is wet. These joints can be made above or below grade.

Schedule 80 or heavier pipe can be threaded with dies specifically designed for plastic piping. An approved thread lubricant or tape must be applied to the male threads only.

Asbestos-cement pipe is joined by a sleeve coupling of the same composition as the pipe and sealed with an elastomeric ring.

Brazed joints must be made on clean surfaces with an approved flux. The filler used for brazing must be an approved material. The same is true for welded joints.

Caulked joints are rarely used today. Joints for hub and spigot pipe must be firmly packed with oakum or hemp. Molten lead is poured in one operation to a depth of not less than 1 inch. The lead must not recede more than 0.125 inch below the rim of the hub and must be caulked tight. The lead joint cannot be painted, varnished, or otherwise coated until after the joint is tested and approved. When a lead joint is made, the joint is to be made in one pouring and caulked tight. Acid-resistant rope and acid-proof cement are permitted.

Compression gasket joints must be compressed when the pipe is fully inserted. Joints between concrete pipe and fittings are to be made with an elastomeric seal. Mechanical joints on drainage pipe must be made with an elastomeric seal. It is not permissible to install mechanical joints in above-grade systems unless otherwise approved. When making joints with stainless- steel drainage systems and other types of systems, a mechanical joint must be used. If an O-ring is used with a stainless-steel drainage system, you must use an elastomeric seal.

SUPPORT

How you support your pipes is also regulated by the plumbing code. There are requirements for the type of materials you may use and how they may be used. Let's see what they are.

The hangers used must be compatible with the pipe they are supporting. You must use a hanger that will not have a detrimental effect on your piping. For example, you may not use galvanized straphanger to support copper pipe. As a rule of thumb, the hangers used to support a pipe should be made from the same material as the pipe being supported. For example, copper pipe should be hung with copper hangers. This eliminates the risk of a corrosive action between two different types of materials. If you are using a plastic or plastic-coated hanger, you may use it with all types of pipe. The exception to this rule is for pipes carrying a liquid with a temperature that might affect or melt the plastic hanger.

The hangers used to support pipe must be capable of supporting the pipe at all times. The hanger must be attached to the pipe and to the member holding the hanger in a satisfactory manner. For example, it would not be acceptable to wrap a piece of wire around a pipe and then wrap the wire around the bridging between two floor joists. Hangers should be securely attached to the members supporting them. For example, a hanger should be attached to the pipe and then nailed to a floor joist. The nails used to hold a hanger in place should be made of the same material as the hanger if corrosive action is a possibility.

CODE UPDATE

Heat-fusion joints for polyvinylidene fluoride pipe and tubing joints shall be installed with socket-type heat-fused polyvinylidene fluoride fittings or electrofusion polyvinylidene fittings and couplings.

Both horizontal and vertical pipes require support. The intervals between supports will vary, depending upon the type of pipe being used and whether it is installed vertically or horizontally. The following examples will show you how often you must support the various types of pipes when they are hung horizontally:

- ABS:: every 4 feet
- Cast iron: every 5 feet
- Galvanized: every 12 feet
- PVC: every 4 feet
- DWV copper: every 10 feet.

When these same types of pipes are installed vertically, they must be supported at no less than the following intervals:

- ABS: every 4 feet
- Cast iron: every 15 feet
- Galvanized: every 15 feet
- PVC: every 4 feet
- DWV copper: every 10 feet.

When installing cast-iron stacks, the base of each stack must be supported. This is due to the weight of cast-iron pipe. When installing pipe with flexible couplings, bands, or unions, the pipe must be installed and supported to prevent these flexible connections from moving. In pipes larger than 4 inches in diameter, all flexible couplings must be supported to prevent the force of the pipe's flow from loosening the connection at changes in direction.

Fast Fact

You may not reduce the size of a drainage pipe as it heads for the waste-disposal site.

Pipe-Size Reduction

As mentioned earlier, you may not reduce the size of a drainage pipe as it heads for the waste-disposal site. The pipe size may be enlarged, but it may not be reduced. There is one exception to this rule: Reducing closet bends are allowed.

More Facts to Remember

A drainage pipe installed underground must have a minimum diameter of 2 inches. When you are installing a horizontal branch fitting near the base of a stack, keep the branch fitting away from the point where the vertical stack turns to a horizontal run. The branch fitting should be installed at least 30 inches back on a 3-inch pipe and 40 inches back on a 4-inch pipe. By multiplying the size of the pipe by a factor of ten, you can determine how far back the branch fitting should be installed.

All drainage piping must be protected from the effects of flooding. When leaving a stub of pipe to connect with fixtures planned for the future, the stub must not be more than 2 feet in length and it must be capped. Some exceptions are possible for the prescribed length of a pipe stub. If you need a longer stub, consult your local code officer. Cleanout extensions are not affected by the 2-foot rule.

Multiple buildings situated on the same building lot may not share a common building sewer that connects to a public sewer. Horizontal branches that connect to the bases of stacks must connect at a point not less than ten pipe diameters downstream from the stack. Unless otherwise provided, horizontal branches must connect to horizontal stack offsets at a point located not less than ten pipe diameters downstream from the upper stack.

Horizontal branches connecting to stacks within 2 feet above or below a vertical stack offset, when the offset is located more than four branch intervals below the top of a stack, require that the offset be vented. Vents for vertical offsets are not required where the stack and its offset are sized as a building drain.

Horizontal branches may not connect to a horizontal stack offset or within 2 feet above or below the offset when the offset is located more than four branch intervals below the top of the stack. A vent is required for a stack with a horizontal offset that is located more than four branch intervals below the top of a stack.

If a vertical offset occurs in a soil or waste stack below the lowest horizontal branch, a change in diameter of the stack because of the offset is not required. If a horizontal offset occurs in a soil or waste stack below the lowest horizontal branch, the diameter of the offset and the stack below it must be sized as a building drain.

Drainage pipe installed in food-service areas must not be installed above any working, storage, or eating surfaces when the drainage or waste piping is exposed. Obviously, this is to protect food areas from contaminants that might be associated with exposed drainage or waste piping.

Lead bends and traps must have a minimum wall thickness of 0.125 inch.

Mechanical joints are not allowed for use on drainage pipe that is installed aboveground unless special permission is obtained. When mechanical joints are used, they must be made with an approved elastomeric seal. All mechanical joints are to be made in accordance with instructions provided by the product manufacturer. When joining pipes of different types of material, mechanical joints must be used. The mechanical joint may be either a compression or mechanical-sealing type of device.

Joints used for glass pipe must be made with a TFE seal. These joints are to be made in accordance with the manufacturer's instructions.

When a pump is used for a drainage system, a check valve and a full-open valve are required. These devices must be installed so that they are accessible. The full-open valve is to be installed in the discharge piping on the discharge side of the check valve. When possible, the valves are to be installed above the sump cover. If the discharge piping is below grade, the valves

CODE UPDATE

Clear-water waste receptors are assigned a rating of one fixture unit.

must be installed outside the sump in an access pit so that the valves are accessible. There is an exception to this rule: In one- and two-family dwellings, a full-open valve is not required. Only a check valve is required when occupancy is limited to one or two families.

FITTINGS

Fittings are also a part of the drainage system. Knowing when, where, and how to use the proper fittings is mandatory for the installation of a drainage system. Fittings are used to make branches and to change direction.

When you wish to change direction with a pipe, it can change from a horizontal run to a vertical rise. You may be going from a vertical position to a horizontal one, or you might only want to offset the pipe in a horizontal run. Each of these three categories requires the use of different fittings. Let's take each circumstance and examine the fittings allowed.

Offsets

When you want to change the direction of a horizontal pipe, you must use fittings approved for that purpose. Those choices are:

- Sixteenth bend
- Eighth bend
- Sixth bend
- Long-sweep fitting
- Combination wye-and-eighth bend
- Wye.

Any of these fittings are generally approved for changing direction with horizontal piping, but, as always, it is best to check with your local code officer for current regulations.

Horizontal to Vertical Changes Of Direction

You have a wider range of choice in selecting a fitting for changing from a horizontal position to a vertical position. There are nine possible candidates available for this type of change in direction. The choices are:

- Sixteenth bend
- Eighth bend
- Sixth bend
- Long-sweep fitting
- Combination wye-and-eighth bend
- Wye
- Quarter bend
- Short-sweep fitting
- Sanitary tee.

You may not use a double sanitary tee in a back-to-back situation if the fixtures being served are of a blowout or pump type. Double sanitary tees must not be used to receive the waste of back-to-back water closets. For example, you could not use a double sanitary tee to receive the discharge of two washing machines if the machines were positioned back-to-back. The sanitary tee's throat is not deep enough to keep drainage from feeding back and forth between the fittings. In a case like this, use a double combination wye-and-eighth bend. The combination fitting has a much longer throat and will prohibit wastewater from transferring across the fitting to the other fixture. There is an exception to this rule: A double sanitary tee can be used to accept the waste from a back-to-back water-closet connection when the horizontal developed length between the outlet of the water closet and the connection to the double sanitary tee is 18 inches or greater.

Vertical to Horizontal Changes in Direction

There are seven fittings allowed to change direction from vertical to horizontal. These fittings are:

- Sixteenth bend
- Eighth bend
- Sixth bend
- Long-sweep fitting
- Combination wye-and-eighth bend
- Wye
- Short sweep fitting 3 inches or larger.

Some codes prohibit a fixture-outlet connection within 8 feet of a vertical to horizontal change in direction of a stack if the stack serves a suds-producing fixture. A suds-producing fixture could be a laundry fixture, a dishwasher, a bathing unit, or a kitchen sink. This rule does not apply to single-family homes and stacks in buildings with less than three stories.

CHAPTER

8

Indirect and Special Wastes

Indirect-waste requirements can pertain to a number of types of plumbing fixtures and equipment. These might include a clothes-washer drain, a condensate line, a sink drain, or the blowoff pipe from a relief valve, just to name a few. These indirect wastes are piped in this manner to prevent the possibility of contaminated matter backing up the drain into a potable-water or food source, among other things.

Most indirect-waste receptors are trapped. If the drain from the fixture is more than 2 feet long, the indirect-waste receptor must be trapped. However, this trap rule applies to fixtures such as sinks, not to an item such as a blowoff pipe from a relief valve. In some areas a drain that is less than 5 feet long does not have to be trapped. If a floor drain is located within an area subject to freezing, the waste line serving the drain must not be trapped and must indirectly discharge into a waste receptor located outside the area subject to freezing.

The safest method of indirect-waste disposal is accomplished by using an air gap. When an air gap is used, the drain from the fixture terminates

Fast Fact

If the drain from the fixture is more than 2 feet long, the indirect-waste receptor must be trapped.

above the indirect-waste receptor, with open-air space between the waste receptor and the drain. This prevents any backup or backsiphonage.

Some fixtures, depending on local code requirements, may be piped with an air break rather than an air gap. With an air break, the drain may extend below the flood-level rim and terminate just above the trap's seal. The risk of an air break is the possibility of a backup. Since the drain is run below the flood-level rim of the waste receptor, it is possible that the waste receptor could overflow and back up into the drain. This could create contamination; in cases where contamination is likely, an air gap will be required. Check with your local code office before using an air break.

Domestic dishwashing machines must discharge indirectly through an air gap or air break into a standpipe or other approved receptor. One such approved receptor is the tailpiece of a kitchen sink when the waste line from the dishwasher is connected to and discharging through an air gap and wye-branch fitting. It is also acceptable for a garbage disposer to receive the waste that has passed through an air gap.

Standpipes, such as those used for washing machines, are a form of indirect-waste receptor. A standpipe used for this purpose in most jurisdictions must extend at least 18 inches but not more than 42 inches above the trap seal. Standpipes are to be individually trapped and accessible. If a clear-water waste receptor is located in a floor, some codes require the lip

Trade Tip

The safest method of indirect-waste disposal is accomplished by using an air gap.

Fast Fact

Domestic dishwashing machines must discharge indirectly through an air gap or air break into a standpipe or other approved receptor.

of the receptor to extend at least 2 inches above the floor. This prevents the waste receptor from being used as a floor drain.

The standpipe for an automatic clothes washer must have a minimum diameter of 2 inches. The fixture drain must connect to a branch drain or drainage stack that has a minimum diameter of 3 inches.

Choosing the proper size for a waste receptor is generally based on the receptor's ability to handle the discharge from a drain without excessive splashing. If you are concerned with sizing a particular waste receptor, ask your local code officer for a ruling.

Buildings used for food preparation, storage, and similar activities are required to have their fixtures and equipment discharge drainage through an air gap. Dishwashers and open culinary sinks are sometimes exceptions. Some code regions require that a discharge pipe terminate at least 2 inches above the receptor. Other regions require the distance to be a minimum of 1 inch. You may find that your local code requires the air-gap distance to be a minimum of twice the size of the pipe discharging the waste. For example, a 0.5-inch discharge pipe would require a 1-inch air gap. Check your local code requirements closely on this matter.

Floor drains located within walk-in refrigerators or freezers in food-service and food establishments must be indirectly connected to the sanitary

Fast Fact

Buildings used for food preparation, storage, and similar activities are required to have their fixtures and equipment discharge drainage through an air gap.

Fast Fact

Most codes prohibit the installation of an indirect-waste receptor in any room containing toilet facilities. There can be an exception: the installation of a receptor for a clothes washer when it is installed in the same room.

drainage system by means of an air gap. There is an exception to this rule: Where protected against backflow by a backwater valve, such floor drains can be indirectly connected to the sanitary drainage system by means of an air break or an air gap. Waste receptors are permitted in the form of a hub or pipe extending not less than 1 inch above a water-impervious floor and are not required to have a strainer.

Most codes prohibit the installation of an indirect-waste receptor in any room containing toilet facilities. There can be an exception: the installation of a receptor for a clothes washer when it is installed in the same room. Indirect-waste receptors may not be installed in closets and other unvented areas. Indirect-waste receptors must be accessible. Code generally requires all receptors to be equipped with a means of preventing solids with diameters of 0.5 inch or larger from entering the drainage system. These straining devices must be removable to allow for cleaning.

When you are dealing with extreme water temperatures in wastewater, such as with a commercial dishwasher, the drain must be piped to an indirect waste. The indirect waste will be connected to the sanitary plumbing system, but the dishwasher drain may not connect to the sanitary system directly if the wastewater temperature exceeds 140°F. The discharge from

CODE UPDATE

Sinks used for cleaning pots, pans, dishes, and service ware for the serving of food must discharge indirectly through an air gap or an air break or directly connect to a drainage system.

Trade Tip

When you are dealing with extreme water temperatures in wastewater, such as with a commercial dishwasher, the drain must be piped to an indirect waste.

a commercial dishwasher must pass through an air gap or air break and enter a standpipe or approved waste receptor. Steam pipes may not be connected directly to a sanitary drainage system. Local regulations may require the use of special piping, sumps, or condensers to accept high-temperature water. The direct connection of any dishwasher to the sanitary drainage system is likely to be prohibited.

Clear-water waste from a potable source must be piped to indirect waste through an air gap. Sterilizers and swimming pools might provide two examples of when this rule would be used. Clear water from nonpotable sources, such as a drip from a piece of equipment, must be piped to an indirect-waste receptor. Some jurisdictions allow an air break in place of an air gap. Other code regions require any waste entering the sanitary drainage system from an air conditioner to do so through an indirect waste.

Where wastewater from swimming pools, backwash from filters, and water from pool deck drains discharge to the building drainage system, the discharge must be through an indirect-waste pipe by means of an air gap.

SPECIAL WASTES

Special wastes are wastes that may have a harmful effect on a plumbing or waste-disposal system. Possible locations for special-waste piping might

Fast Fact

Clear-water waste from a potable source must be piped to indirect waste through an air gap.

Trade Tip

Special wastes are wastes that may have a harmful effect on a plumbing or waste-disposal system.

include photographic labs, hospitals, or buildings where chemicals or other potentially dangerous wastes are dispersed. Small, personal-type photo darkrooms do not generally fall under the scrutiny of these regulations. Buildings that are considered to have a need for special-waste plumbing are often required to have two plumbing systems, one system for normal sanitary discharge and a separate system for the special wastes. Before any special wastes are allowed to enter a sanitary drainage system, they must be neutralized, diluted, or otherwise treated.

Depending upon the nature of the special wastes, special materials may be required. When you venture into the plumbing of special wastes, it is always best to consult the local code officer before proceeding with your work.

CHAPTER

9

Vents

Most people don't think much about vents when they consider the plumbing in their home or office, but vents play a vital role in the scheme of sanitary plumbing. Many plumbers underestimate the importance of vents. The sizing and installation of vents often cause more confusion than the same tasks applied to drains. This chapter will teach you the role and importance of vents. It will also instruct you in the proper methods of sizing and installing vents.

Whether you are working with simple individual vents or complex island vents, this chapter will improve your understanding of their installation. Why do we need vents? Vents perform three easily identified functions. The most obvious function of a vent is its capacity to carry sewer gas out of a building and into the open air. A less obvious but equally important aspect of the vent is its ability to protect the seal in the trap it serves. The third characteristic of the vent is its ability to enable drains to drain faster and better. Let's look more closely at each of these factors.

Fast Fact

Good trap seals are essential to sanitary plumbing systems.

SEWER GAS

Vents transport sewer gas through a building to an open-air space without exposing occupants of the building to the gas. Why is this important? Sewer gas can cause health problems. The effect of sewer gas on individuals will vary, but it should be avoided by all. In addition to health problems caused by sewer gas, explosions are also possible when sewer gas is concentrated in a poorly ventilated area. Yes, sewer gas can create an explosion when it is concentrated, confined, and ignited. As you can see just for this reason alone vents are an important element of a plumbing system.

PROTECTING TRAP SEALS

Another job plumbing vents perform is the protection of trap seals. The water sitting in a fixture's trap blocks the path of sewer gas trying to enter the plumbing fixture. Without a trap seal, sewer gas could rise through the drainage pipe and enter a building through a plumbing fixture. As mentioned above, this could result in health problems and the risk of explosion. Good trap seals are essential to sanitary plumbing systems.

Vents protect trap seals. How do they do it? They regulate the atmospheric pressure applied to the seals. It is possible for pressures to rise in unvented traps to a point where the contents actually expel into the fixture. This is

Trade Tip

Vents protect trap seals. How do they do it? They regulate the atmospheric pressure applied to the seals.

Trade Tip

Vents help fixtures to drain faster. The air allowed in by the vent keeps the water moving at a more rapid pace.

not a common problem, but if it occurs, the plumbing fixture could become contaminated.

A more likely problem is when the pressure on a trap seal is reduced to a near vacuum. When this happens, the water creating the trap seal is sucked out of the trap and down the drain. Once the water is taken from the trap, there is no trap seal. The trap will remain unsealed until water is replaced in the trap. Without water in it, a trap is all but useless. Vents prevent these extreme atmospheric pressure changes, thus protecting the trap seal.

Air-admittance valves must be sized in accordance with the standard for the size of the vent to which the valve is connected. The design of a vent system can be created with an approved computer-program method. Capacity requirements for a vent system must be based on the air capacity requirements of the drainage system under a peak load condition.

DRAINAGE

Have you ever drained your sink or bathtub and watched the tiny water tornados? When you see the fast swirling action of water being pulled down a drain, it usually indicates that the drain is well vented. If water is sluggish and moves out of the fixture like a lazy river, the vent for the fixture, if there is one, is not performing at its best.

Vents help fixtures to drain faster. The air allowed in from the vent keeps the water moving at a more rapid pace. This not only entertains us with tiny tornados, but it aids in the prevention of clogged pipes. It is possible for drains to drain too quickly, removing the liquids and leaving hair, grease, and other potential pipe blockers present. However, if a pipe is properly graded and does not contain extreme vertical drops into improper fittings, such problems should not occur.

VENT EXCEPTIONS

Most local plumbing codes require all fixture traps to be vented, but there are exceptions. In some jurisdictions, combination waste and vent systems are used. In a combination waste and vent system, vertical vents are rare. Instead of vertical vents, larger drainage pipes are used. The larger diameter of the drain allows air to circulate in the pipe, eliminating the need for a vent as far as satisfactory drainage is concerned. I have worked with both types of systems, predominantly vented systems, and in my opinion, vented systems perform much better than combination waste and vent systems.

Combination waste and vent systems do not have vents on each fixture, so how is the trap seal protected? Trap seals in a combination system are protected through the use of antisyphon or drum traps. Vented systems normally use P-traps. By using an antisiphon or drum, the trap is not susceptible to backsiphonage. Since these traps are larger, deeper, and made so that the water is not replaced with each use of the fixture, they are not required to be vented, subject to local code requirements. Most jurisdictions prohibit the use of drum traps and require traps to be vented. Before you install your plumbing, check with the local code officer for the facts pertinent to your location. Fittings for vent piping must be compatible with the piping used.

A combination drain and vent system can serve only the following types of fixtures:

- Floor drains
- Sinks
- Lavatories
- Drinking fountains.

Fast Fact

Combination waste and vent systems are not allowed with garbage disposers.

Combination waste and vent systems are not allowed with garbage disposers. The only vertical pipe of a combination drain and vent system is the connection between the fixture drain of a sink, lavatory, or drinking fountain and the horizontal combination drain and vent pipe. The maximum vertical distance is 8 feet.

INDIVIDUAL VENTS

Individual vents are, as the name implies, vents that serve individual fixtures. These vents only vent one fixture, but they may connect into another vent that extends to the open air. Individual vents do not have to extend from the fixture being served to the outside air without joining another part of the venting system, but they must eventually vent to open-air space.

Sizing an individual vent is easy. The vent must be at least one-half the size of the drain it serves, but it may not have a diameter of less than 1.25 inches. For example, a vent for a 3-inch drain could, in most cases, have a diameter of 1.5 inches. A vent for a 1.5-inch drain may not have a diameter of less than 1.25 inches.

RELIEF VENTS

Relief vents are used in conjunction with other vents. Their purpose is to provide additional air to the drainage system when the primary vent is too far from the fixture. Relief vents must be at least one-half the size of the pipes they are venting. For example, if a relief vent is venting a 3-inch pipe, the relief vent must have a 1.5-inch or larger diameter. Use the sizing tables in your local code book to establish minimum size requirements. Relief vents may be used to vent more than one fixture.

Fast Fact

Relief vents must be at least one-half the size of the pipes they are venting.

When relief vents are required on stacks of more than ten branch intervals, the lower end of each relief vent must connect to the soil or waste stack through a wye below the horizontal branch serving the floor, and the upper end must connect to the vent stack through a wye not less than 3 feet above the floor.

CIRCUIT VENTS

Circuit vents are used with a battery of plumbing fixtures. Circuit vents are normally installed just before the last fixture of the battery. Then the circuit vent is extended upward to the open air or tied into another vent that extends to the outside. Circuit vents may tie into stack vents or vent stacks. When sizing a circuit vent, you must account for its developed length. But, in any event, the diameter of a circuit vent must be at least one-half the size of the drain it is serving.

VENT SIZING

What effect does the length of the vent have on the vent's size? The developed length, the total linear footage of pipe making up the vent, is used in conjunction with factors provided in code books to determine vent sizes. To size circuit vents, branch vents, and individual vents for horizontal drains, you must use this method of sizing.

The criteria needed for sizing a vent based on developed length are: the grade of the drainage pipe, the size of the drainage pipe, the developed length of the vent, and the factors allowed by local code requirements. Knowing this information, you will use the sizing tables in your local code book to establish pipe sizing.

Fast Fact

A branch vent that has a developed length in excess of 40 feet must increase the pipe sizing by one nominal size for the entire developed length of the vent.

BRANCH VENTS

Branch vents are vents extending horizontally and connecting multiple vents together.

Branch vents are sized with the developed-length method. A branch or individual vent that is the same size as the drain it serves is unlimited in the developed length it may reach. A branch vent that has a developed length in excess of 40 feet must increase the pipe sizing by one nominal size for the entire developed length of the vent. Be advised that not all local codes use the same sizing charts, so check your local code before you trust your sizing.

VENT STACKS

A vent stack is a pipe used only for the purpose of venting. Vent stacks extend upward from the drainage piping to the open air outside a building. Vent stacks are used as connection points for other vents, such as branch vents. A vent stack is a primary vent that accepts the connection of other vents and vents an entire system. Vent stacks run vertically and are sized a little differently.

The basic procedure for sizing a vent stack is similar to that used with branch vents, but there are some differences. You must know the size of the soil stack, the number of fixture units carried by the soil stack, and the developed length of the vent stack. With this information and the regulations of your local plumbing code, you can size your vent stack. The same sizing method is used when computing the size of stack vents.

STACK VENTS

Stack vents are really two pipes in one. The lower portion of the pipe is a soil pipe, and the upper portion is a vent. This is the type of primary vent most

Fast Fact

Offsets are permitted in the stack vent and must be located at least 6 inches above the flood level of the highest fixture.

often found in residential plumbing. Stack vents are sized with the same methods used for vent stacks. Offsets are permitted in the stack vent and must be located at least 6 inches above the flood level of the highest fixture.

COMMON VENTS

Common vents are single vents that vent multiple traps. Common vents are only allowed when the fixtures being served by the single vent are on the same floor level. Some jurisdictions require the drainage of fixtures being vented with a common vent to enter the drainage system at the same level. Normally, not more than two traps can share a common vent, but there is an exception in some regions. In some areas you may vent the traps of up to three lavatories with a single common vent. Common vents are sized with the same technique applied to individual vents.

ISLAND VENTS

Island vents are unusual-looking vents. They are allowed for sinks and lavatories. The primary use for these vents is with the trap of a kitchen sink when the sink is placed in an island cabinet. The vent must rise as high as possible under the cabinet before it takes a U-turn and heads back downward. Since this piping does not rise above the flood-level rim of the fixture, it must be considered a drain. Fittings approved for drainage must be used in making an island vent. The vent portion of an island vent must be equipped with a cleanout. The vent may not tie into a regular vent until it rises at least 6 inches above the flood-level rim of the fixture.

WET VENTS

Wet vents are pipes that serve as a vent for one fixture and a drain for another. Only the fixtures within a bathroom group may connect to a wet-vented horizontal branch drain. Additional fixtures must discharge down-

Trade Tip

Island vents are allowed for sinks and lavatories.

Trade Tip

By effectively using wet vents you can reduce the amount of pipe, fittings, and labor required to vent a bathroom group or two.

stream of the wet vent. Wet vents, once you know how to use them, can save you a lot of money and time. By effectively using wet vents you can reduce the amount of pipe, fittings, and labor required to vent a bathroom group or two. Dry vents connected to wet vents must be sized based on the largest required diameter of pipe within the wet-vent system served by the dry vent.

Any combination of fixtures within two bathroom groups located on the same floor level can be vented by a vertical wet vent. This type of vent is required to extend from the connection to the dry vent down to the lowest fixture drain connection. Each fixture is required to connect independently to the vertical wet vent. When water closets are connected to this type of system, the connections must be made at the same elevation. Fixtures other than water closets are to connect to the system either at the same level as the water closets or above those connections. The dry-vent connection to the vertical wet vent has to be an individual or common vent serving one or two fixtures.

The sizing of wet vents is based on fixture units. The size of the pipe is determined by how many fixture units it may be required to carry. A 3-inch wet vent can handle twelve fixture units. A 2-inch wet vent is rated for four fixture units, and a 1.5-inch wet vent is allowed only one fixture unit. It is acceptable to wet-vent two bathroom groups, six fixtures, with a single vent, but the bathroom groups must be on the same floor level.

Trade Tip

The sizing of wet vents is based on fixture-units.

Depending upon local regulations, the horizontal branch connecting to the drainage stack may have to enter at a level equal to or below the water-closet drain. However, the branch may connect to the drainage at the closet bend. When wet-venting two bathroom groups, the wet vent must have a minimum diameter of 2 inches. Kitchen sinks and washing machines may not be drained into a 2-inch combination waste and vent. Water closets and urinals are restricted to vertical combination waste and vent systems.

If wet venting is allowed on different floor levels in your region, the vents must have at least a 2-inch diameter. Water closets that are not located on the highest floor must be back-vented. If, however, the wet vent is connected directly to the closet bend with a 45-degree bend, the toilet being connected is not required to be back-vented even if it is on a lower floor.

Wet venting in some regions may be limited to vertical piping. These vertical pipes are restricted to receiving the waste from fixtures that have fixture-unit ratings of two or less and that serve to vent no more than four fixtures. Wet vents must be one pipe size larger than normally required, but they must never be smaller than 2 inches in diameter.

CROWN VENTS

A crown vent is a vent that extends upward from a trap or trap arm. Crown-vented traps are not allowed. When crown vents are used, they are normally used on trap arms, but even then they are not common. The vent must be on the trap arm, and it must be behind the trap by a distance equal to twice the pipe size. For example, on a 1.5-inch trap, the crown vent would have to be 3 inches behind the trap on the trap arm.

Trade Tip

If you will be installing a pneumatic sewer ejector, you will need to run the sump vent to outside air without tying it into the venting system when using a standard sanitary plumbing system.

Fast Fact

Any building equipped with plumbing must also be equipped with a main vent. The size of this vent must be no less than one-half the size of the building drain.

VENTS FOR SUMPS AND SEWER PUMPS

When sumps and sewer pumps are used to store and remove sanitary waste, the sump must be vented. If you will be installing a pneumatic sewer ejector, you will need to run the sump vent to outside air without tying it into the venting system when using a standard sanitary plumbing system. If your sump will be equipped with a regular sewer pump, you may tie the vent from the sump back into the main venting system for the other sanitary plumbing.

Additional rulings apply in some regions. You may find that sump vents must not be smaller than 1.25-inch pipe. The size requirements for sump vents are determined by the discharge of the pump. For example, a sewer pump capable of producing 20 gallons a minute could have its sump vented for an unlimited distance with a 1.5-inch pipe. If the pump was capable of producing 60 gallons per minute, a 1.5-inch pipe could not have a developed length of more than 75 feet.

In most cases, a 2-inch vent is used on sumps, and the distance allowed for developed length is not a problem. However, if your pump will pump more than 100 gallons per minute, you had better take the time to do some math. Your code book will provide you with the factors you need to size your vent, and the sizing is easy. You simply look for the maximum discharge capacity of your pump and match it with a vent that allows the developed length you need. This concludes the general description and sizing techniques for various vents. Next we are going to look at regulations dealing with the installation methods for vents.

INSTALLATION REQUIREMENTS

Since there are so many types of vents and their role in the plumbing system is so important, there are many regulations affecting their installation.

What follows are specifics for installing various types of vents.

Main Vents

Any building equipped with plumbing must also be equipped with a main vent. The size of this vent must be no less than one-half the size of the building drain. This vent must run undiminished in size and as directly as possible from the building drain through to the open air or to a vent header that extends to the open air. Any plumbing system that receives the discharge from a water closet must have either a main vent stack or stack vent. This vent must originate at a 3-inch drainage pipe and extend upward until it penetrates the roof of the building and meets outside air. The vent size requirements call for a minimum diameter of 3 inches. However, some codes allow the main stack in detached buildings, where the only plumbing is a washing machine or laundry tub, to have a diameter of 1.5 inches.

Main vents that are vent stacks must connect to building drains or to the bases of drainage stacks in compliance with the plumbing code. A main vent that is a stack vent must be an extension of the drainage stack. When a vent stack connects to a building drain, the connection is to be located downstream of the drainage stack and within a distance of ten times the diameter of the drainage stack.

Multiple Branch Vents

Multiple branch vents that exceed 40 feet in developed length must be increased by one nominal size for the entire developed length of the vent pipe. When a vent penetrates a roof, it must be flashed or sealed to prevent water from leaking past the pipe and through the roof. Metal flashings with rubber collars are normally used for flashing vents, but more modern flashings are made from plastic rather than metal.

The vent must extend above the roof to a certain height. The height may fluctuate among geographical locations. Average vent extensions are between 12 and 24 inches, but check with your local regulations to determine the minimum height in your area.

When vents terminate in the open air, the proximity of their location to windows, doors, or other ventilating openings must be considered. If a vent were placed too close to a window, sewer gas might be drawn into the building when the window was open. Vents should be kept 10 feet from any window, door, opening, or ventilation device. If the vent cannot be kept at least 10 feet from the opening, it should extend at least 2 feet above the opening. Depending upon your local region, the vent may be required to extend at least 3 feet above the opening.

If the roof being penetrated by a vent is used for activities other than weather protection, such as a patio, the vent must extend several feet above the roof. Some regions require the vent to extend at least 5 feet above the roof. Other regions require the extension to rise even higher. In cold climates, vents must be protected from freezing. Condensation can collect on the inside of vent pipes. In cold climates this condensation may turn to ice. As the ice mass grows, the vent becomes blocked and useless.

This type of protection is usually accomplished by increasing the size of the vent pipe. This ruling normally applies only in areas where temperatures are expected to be below 0°F. Some codes require vents in this category to have a minimum diameter of 3 inches. If this requires an increase in pipe size, the increase must be made at least 1 foot below the roof. In the case of side-wall vents, the change must be made at least 1 foot inside the wall. In some regions, all vents must have diameters of at least 2 inches but never less than the normally required vent size. Any change in pipe size must take place at least 12 inches before the vent penetrates into open air, and the vent must extend to a height of 10 inches.

Side-Wall Vents

There may be occasions when it is better to terminate a plumbing vent out the side of a wall rather than through a roof. Some jurisdictions don't allow this practice, but others do. Some regions prohibit side-wall vents from terminating under any building's overhang. When side-wall vents are installed, they must be protected against birds and rodents with a wire

CODE UPDATE

Air admittance valves without an engineered design shall not be utilized to vent sumps or tanks of any type.

mesh or similar cover. Side-wall vents must not extend closer than 10 feet to the property boundary of the building lot. If the building is equipped with soffit vents, side-wall vents may not be used in such a way that they terminate under the soffit vents. This rule is in effect to prevent sewer gas from being sucked into the attic of the home.

Vent Stacks

Some codes require buildings that have soil stacks with more than five branch intervals to be equipped with a vent stack. Others require a vent stack in buildings that have at least ten stories above the building drain. The vent stack will normally run up near the soil stack. The vent stack must connect into the building drain at or below the lowest branch interval. The vent stack must be sized according to tables in your local code book. The vent stack may be required to be connected at an interval within ten times its pipe size on the downward side of the soil stack. This means that a 3-inch vent stack must be within 30 inches of the soil stack on the downward side of the building drain.

Stack Vents

Check your local code to see if stack vents must be connected to the drainage stack at intervals of every five stories. If so, the connection must be made with a relief yoke vent. The yoke vent must be at least as large as either the vent stack or soil stack, whichever is smaller. This connection must be made with a wye fitting at least 42 inches off the floor.

In large plumbing jobs where there are numerous branch intervals, it may be necessary to vent offsets in the soil stack. Normally, the offset must be more than 45 degrees to warrant an offset vent. It is common for offset vents to be required when the soil-stack offsets have five or more branch intervals above them.

Fast Fact

In large plumbing jobs where there are numerous branch intervals, it may be necessary to vent offsets in the soil stack.

Fast Fact

Most vents can be tied into other vents, such as a vent stack or stack vent. But the connection for the tie-in must be at least 6 inches above the flood-level rim of the highest fixture served by the vent.

Just as drains are installed with a downward pitch, vents must also be installed with a consistent grade. Vents should be graded to allow any water entering the vent pipe to drain into the drainage system. A typical grade for vent piping is 0.25 inch to the foot.

Dry Vents

Dry vents must be installed in a manner to prevent clogging and blockages. You may not lay a fitting on its side and use a quarter bend to turn the vent up vertically. Dry vents should leave the drainage pipe in a vertical position. An easy way to remember this is that if you need an elbow to get the vent up from the drainage, you are doing it wrong.

Most vents can be tied into other vents, such as a vent stack or stack vent. But the connection for the tie-in must be at least 6 inches above the flood-level rim of the highest fixture served by the vent.

Circuit Vents

Some regions allow the use of circuit vents to vent fixtures in a battery. The drain serving the battery must be operating at one-half its fixture-unit rating. If the application is on a lower-floor battery with a minimum of three fixtures, relief vents are required. You must also pay attention to the fixtures draining above these lower-floor batteries.

Trade Tip

Circuit vents may at times be used to vent up to eight fixtures utilizing a common horizontal drain.

When a fixture with a fixture-rating of four or less and a maximum drain size of 2 inches is above the battery, every vertical branch must have a continuous vent. If a fixture with a fixture-unit rating exceeding four is present, all fixtures in the battery must be individually vented. Circuit-vented batteries may not receive the drainage from fixtures on a higher level.

Circuit vents should rise vertically from the drainage. However, the vent can be taken off the drainage horizontally if the vent is washed by a fixture with a rating of no more than four fixture-units. The washing cannot come from a water closet. The pipe being washed must be at least as large as the horizontal drainage pipe that it is venting.

Circuit vents may at times be used to vent up to eight fixtures utilizing a common horizontal drain. Circuit vents must be dry vents, and they should connect to the horizontal drain in front of the last fixture on the branch. The horizontal drain being circuit-vented must not have a grade of more than 1 inch per foot. Some code requirements interpret the horizontal section of drainage being circuit-vented as a vent. If a circuit vent is venting a drain with more than four water closets attached to it, a relief vent must be installed in conjunction with the circuit vent.

Vent placement in relation to the trap it serves is important and regulated. The maximum allowable distance between a trap and its vent will depend on the size of the fixture drain and trap.

All vents, except those for fixtures with integral traps, should connect above the trap seal. A sanitary tee fitting should be used when going from a vertical stack vent to a trap. Other fittings with a longer turn, such as a combination wye-and-eighth bend, will place the trap in more danger of backsiphonage. I know that this goes against the common sense of a smoother flow of water, but the sanitary tee reduces the risk of a vacuum.

All individual, branch, and circuit vents are required to connect to a vent stack, stack vent, or air-admittance valve or to extend to open air. Vents for future-use rough-ins must be not less than one-half the size of the drain to be served. Rough-in vents must be labeled as vents and must either be connected to the vent system or extend to open air.

SUPPORTING YOUR PIPE

Vent pipes must be supported. Vents may not be used to support antennas, flagpoles, and similar items. Depending upon the type of material you are using and whether the pipe is installed horizontally or vertically, the spacing between hangers will vary. Both horizontal and vertical pipes require support. The regulations in the plumbing code apply to the maximum distance between hangers.

Some interceptors, such as those used as a settling tank that discharges through a horizontal indirect waste, are not required to be vented in certain regions. However, the interceptor receiving the discharge from the unvented interceptor must be properly vented and trapped.

Traps for sinks that are a part of a piece of equipment, such as a soda fountain, are not required to be vented when venting is impossible. But these drains must drain through an indirect waste to an approved receptor.

Depending upon your region, you may find that all soil stacks that receive the waste of at least two vented branches must be equipped with a stack vent or a main stack vent. Except when approved, fixture drainage may not be allowed to enter a stack at a point above a vent connection. Side-inlet closet bends are allowed to connect to fixtures that are vented. However, these connections may not be used to vent a bathroom, unless the connection is washed by a fixture. All fixtures dumping into a stack below a higher fixture must be vented, except when special approval is granted for a variance. Stack vents and vent stacks must connect to a common vent header prior to vent termination.

Up to two fixtures, set back-to-back or side-by-side, within the allowable distance between the traps and their vents may be connected to a common

CODE UPDATE

A wet vent extends from the connection with a dry vent along the direction of the flow in the drain pipe to the most downstream fixture drain connection to a horizontal branch drain.

horizontal branch that is vented by a common vertical vent. However, the horizontal branch must be one pipe size larger than normal. When applying this rule, the following ratings apply: shower drains, 3-inch floor drains, 4-inch floor drains, pedestal urinals, and water closets with fixture-unit ratings of four are considered to have 3-inch drains.

Some fixture groups are allowed to be stack-vented without individual back vents. These fixture groups must be located in one-story buildings or on the top floor of the building, with some special provisions. Fixtures located on the top floor must connect independently to the soil stack, and the bathing units and water closets must enter the stack at the same level.

This same stack-venting procedure can be adapted to work with fixtures on lower floors. The stack being stack-vented must enter the main soil stack though a vertical eighth-bend-and-wye combination. The drainage must enter above the eighth bend. A 2-inch vent must be installed on the fixture group. This vent must be 6 inches above the flood-level rim of the highest fixture in the group.

Some fixtures can be served by a horizontal waste that is within a certain distance of a vent. When piped in this manner, bathtubs and showers are both required to have 2-inch P-traps. These drains must run with a minimum grade of 1.25 inches per foot. A single drinking fountain can be rated as a lavatory for this type of piping. With this type of system, fixture drains for lavatories may not exceed 1.25 inches, and sink drains cannot be larger than 1.5 inches in diameter.

In multistory situations, it is possible to drain up to three fixtures into a soil stack above the highest water closet or bathtub connection without reventing. To do this, certain requirements must be met. These requirements are as follows:

- Minimum stack size of 3 inches is required
- Approved fixture-unit load on stack is met
- All lower fixtures must be properly vented
- All individually unvented fixtures are within allowable distances to the main vent
- Fixture openings must not exceed the size of their traps
- All code requirements must be met and approved.

CODE UPDATE

Dry-vent connections for vertical wet-vent systems shall be individual vents or common vents for the most upstream fixture drain.

COMBINATION WASTE AND VENT SYSTEMS

Most jurisdictions limit the types of fixtures can be served by combination waste and vent systems, but not all. In many locations it is a code violation to include a toilet on a combination system, but Maine, for example, has allowed toilets with this type of system. Since combination systems can get you into a sticky situation, you should consult your local code officer before using them. I will, however, explain how this system works in general.

The type of fixtures you are allowed to connect to in a combination waste and vent system may be limited. In some areas the only fixtures allowed on the combination system are: floor drains, standpipes, sinks, and lavatories. Other areas will allow showers, bathtubs, and even toilets to be installed with the combo system. You will have to check your local regulations to see how they affect your choices of plumbing systems.

Combination waste and vent systems are comprised mainly of horizontal piping. Generally, the only vertical pipes are the risers to lavatories, sinks, and standpipes. These pipes may not normally exceed 8 feet in length. This type of system relies on an oversized drainpipe to provide air circulation for drainage. The pipe is often required to be twice the size required for a drain vented normally. The combination system typically must have at least one vent. The vent should connect to a horizontal drainpipe.

A dry vent is required to be connected at any point within the system, or the system can connect to a horizontal drain that is vented according to the plumbing code. Combination drain and vent systems connecting to building drains receiving only the discharge from a stack or stacks must be provided with a dry vent. The vent connection to the combo system must extend at least 6 inches vertically above the flood-level rim of the highest fixture being vented before offsetting horizontally.

Any vertical vent must rise to a point at least 6 inches above the highest fixture being served before it may take a horizontal turn. In a combination system the pipes are rated for fewer fixture units. A 3-inch pipe connecting to a branch or stack may only be allowed to carry twelve fixture units. A 4-inch pipe, under the same conditions, could be restricted to twenty fixture units. Similarly, a 2-inch pipe might only handle three fixture units, and a 1.5-inch pipe may not be allowed. The ratings for these pipes can increase when the pipes are connecting to a building drain.

Stack vents are allowed, but not always in the normal way. All fixtures on a combo system may be required to enter the stack vent individually as opposed to on a branch, as would normally be the case. A stack vent used in a combo system generally must be a straight vertical vent without offsets. The stack vent usually cannot even be offset vertically; it simply cannot be offset. This rule is different in some locations, so check with your local plumbing inspector to see if you are affected by the no-offset rule.

Since stack vents are common and often required in a combination system, you must know how to size these pipes. The sizing is generally done based on the number of fixture-units entering the stack. I will give you an example of how a stack vent for a combo system might be sized.

Since not all pipes run in conjunction with a combination waste and vent system have to follow the combo rules, it is possible that you would have a 1.5-inch pipe entering a stack. The 1.5-inch pipe could only be used if it had an individual vent. It is also possible that the stack vent would be a 1.5-inch pipe.

First, let's look at the maximum number of fixture-units (fu) allowed on a stack:

- 1.5-inch stack: 2 fu
- 2-inch stack: 4 fu
- 3-inch stack: 24 fu
- 4-inch stack: 50 fu
- 5-inch stack: 75 fu
- 6-inch stack: 100 fu.

When you are concerned with the size of a drain dumping into the stack, there are only two pipe sizes to contend with. All pipe sizes larger than 2 inches may dump an unlimited number of fixture-units into the stack. A 1.5-inch pipe may run one fixture-unit into the stack, and a 2-inch pipe may deliver two fixture-units. Sizing your stack is as simple as finding your fixture-unit load on the chart in you local code book. Compare your fixture-units to the chart and select a pipe size rated for that load.

Again, I want to remind you that combination waste and vent systems vary a great deal, so confirm your local requirements before using this type of system.

CHAPTER

10

Traps

We have covered most of the regulations you will need to know about drains and vents. This chapter will round out your knowledge. Here you will learn about traps. Traps have been mentioned before, and you have learned the importance of vents to trap seals, but here you will learn more about traps themselves.

Cleanouts are a necessary part of the drainage system. This chapter will tell you what types of cleanouts you can use and when and where they must be used. Along with cleanouts, backwater valves will be explained. Grease receptors, or grease traps as they are often called, will be explored. By the end of this chapter you should be prepared to tackle just about any DWV job.

CLEANOUTS

What are cleanouts, and why are they needed? Cleanouts are a means of access to the interior of drainage pipes. They are needed so that blockages in drains may be cleared. Without cleanouts, it is much more difficult to snake a drain. In general, the more cleanouts you have, the better. Plumb-

Trade Tip

Generally, cleanouts are required where the building drain meets the building sewer.

ing codes establish minimums for the number of cleanouts required and their placement. Let's look at how these regulations apply to you.

Cleanout Requirements

There are many places in a plumbing system where cleanouts are required. Let's start with sewers. All sewers must have cleanouts. The distances between these cleanouts vary from region to region. Generally, cleanouts are required where the building drain meets the building sewer. The cleanouts may be installed inside or outside the foundation. The cleanout opening must extend upward to the finished floor level or the finished grade outside.

Some jurisdictions prefer that the cleanouts at the junction of building drains and sewers be located outside. If the cleanout is installed inside, it may be required to extend above the flood-level rim of the fixtures served by the horizontal drain. When this is not feasible, allowances may be made. The requirement for a junction cleanout may be waived if there is another cleanout with at least a 3-inch diameter within 10 feet of the junction.

An approved two-way cleanout is allowed in locations where a building drain meets a building sewer. This cleanout is approved for both the building drain and the building sewer.

Once the sewer is begun, cleanouts should be installed every 100 feet. Some regions require cleanouts at an interval distance of 75 feet for 4-inch and

Fast Fact

An approved two-way cleanout is allowed in locations where a building drain meets a building sewer.

Trade Tip

Cleanouts are usually required every time a sewer turns more than 45 degrees.

larger pipe and 50 feet for pipe smaller than 4 inches. Cleanouts are also required in sewers when the pipe changes direction. Cleanouts are usually required every time a sewer turns more than 45 degrees. In some cases, a cleanout is required whenever the change in direction is more than 135 degrees. The general rule for a building sewer is to install a cleanout at intervals that do not exceed 100 feet. This is measured from the upstream entrance of the cleanout. When a building sewer has a diameter of 8 inches or more, the distance between cleanouts can be extended to 200 feet from the junction of the building drain and the building sewer at each change of direction and at intervals not more than 400 feet apart. For these larger sewers, all manholes and manhole covers must be of an approved type.

The cleanouts installed in a sewer must be accessible. This generally means that a standpipe will rise from the sewer to just below ground level. At that point, a cleanout fitting and plug are installed on the standpipe. This allows the sewer to be snaked out from ground level, with little to no digging required.

For building drains and horizontal branches, cleanout location will depend upon pipe size, but they are normally required every 50 feet for pipes with diameters of 4 inches or less. Larger drains may have their cleanouts spaced at 100-foot intervals. Cleanouts are also required on these pipes at each change in direction in excess of 45 degrees. Cleanouts must be installed at the end of all horizontal drain runs. Some jurisdictions do not require cleanouts at intervals less than 100 feet.

Fast Fact

Cleanouts must be installed at the end of all horizontal drain runs.

Cleanout openings must not be used for the installation of new fixtures, except where approved and where another cleanout with equal access and capacity is provided.

As with most rules, there are some exceptions. Some potential exceptions are as follows:

- If a drain is less than 5 feet long and is not used for sinks or urinals, a cleanout is not required.
- A change in direction from a vertical drain with a fifth-bend does not require a cleanout.
- Cleanouts are not required on pipes other than building drains and their horizontal branches that are above the first-floor level.

P-traps and water closets are often allowed to act as cleanouts. When these devices are approved for cleanout purposes, the normally required cleanout fitting and plug at the end of a horizontal pipe run may be eliminated. Not all jurisdictions will accept P-traps and toilets as cleanouts; check your local requirements before omitting standard cleanouts.

Cleanouts must be installed in such a way that the cleanout opening is accessible and allows adequate room for drain cleaning. The cleanout must be installed to go with the flow. This means that when the cleanout plug is removed, a drain-cleaning device should be able to enter the fitting and the flow of the drainage pipe without difficulty.

Cleanouts are frequently required at the base of every stack. This is good procedure at any time, but it is not required by all codes. The height of the cleanout should not exceed 4 feet. Many plumbers install test tees at these locations to plug their stacks for pressure testing. The test tee doubles as a cleanout.

Fast Fact

Cleanouts must be installed in such a way that the cleanout opening is accessible and allows adequate room for drain cleaning.

Fast Fact

Cleanouts are required to be the same size as the pipe they are serving unless the pipe is larger than 4 inches.

When the pipes holding cleanouts will be concealed, the cleanout must be made accessible. For example, if a stack will be concealed by a finished wall, provisions must be made for access to the cleanout. This access could take the form of an access door, or the cleanout could simply extend past the finished wall covering. If the cleanout is serving a pipe concealed by a floor, the cleanout must be brought up to floor level and made accessible. This ruling applies not only to cleanouts installed beneath concrete floors but also to those installed in crawlspaces with very little room to work.

Additional Specifications

There is still more to learn about cleanouts. Size is one of the lessons to be learned. Cleanouts are required to be the same size as the pipe they are serving unless the pipe is larger than 4 inches. If you are installing a 2-inch pipe, you must install 2-inch cleanouts. However, when a P-trap is allowed as a cleanout, it may be smaller than the drain. An example would be a 1.25-inch trap on a 1.5-inch drain. Remember, though, that not all code-enforcement officers will allow P-traps as cleanouts, and they may require the P-trap to be the same size as the drain. Once the pipe size exceeds 4 inches, the cleanouts used should have a minimum size of 4 inches.

When cleanouts are installed, they must provide adequate clearance for drain cleaning. The clearance required for pipes with diameters of 3 inches or more is 18 inches. Smaller pipes require a minimum clearance of 12 inches in front of their cleanouts. Many plumbers fail to remember this regulation. It is common to find cleanouts pointing toward floor joists or too close to walls. You will save yourself time and money by committing these clearance distances to memory.

When a cleanout is installed in a floor, it may be required to have a minimum height clearance of 18 inches and a minimum horizontal clearance of

Fast Fact

Cleanout plugs and plates must be easily removed.

30 inches. No underfloor cleanout is allowed to be placed more than 20 feet from an access opening.

Cleanout plugs and caps must be lubricated with water-insoluble, non-hardening material or tape. Only listed thread tape or lubricants and sealants specifically intended for use with plastics can be used on plastic threads. Conventional pipe-thread compounds, putty, linseed- oil-base products, and unknown lubricants and sealants must not be used on plastic threads.

Acceptable Cleanouts

Cleanout plugs and plates must be easily removed. Access to the interior of the pipe should be available without undue effort or time. Cleanouts can take on many appearances. The U-bend of a P-trap can be considered a cleanout, depending upon local interpretation. A rubber cap, held onto the pipe by a stainless-steel clamp, can serve as a cleanout. The standard female adapter with plug is a fine cleanout. Test tees will work as cleanouts. Special cleanouts, designed to allow the rodding of a drain in either direction, are acceptable.

Cleanouts with plate-style access covers must be fitted with corrosion-resisting fasteners. Plastic cleanout plugs must conform to code requirements. Plugs used for cleanouts are to be constructed of plastic or brass. Countersunk heads are required where raised heads might pose a tripping hazard. Brass cleanout plugs can be used only with metallic drain, waste, and vent piping.

Trade Tip

Plugs used for cleanouts are to be constructed of plastic or brass.

Manholes

The ultimate cleanout is a manhole. You can think of manholes as very big cleanouts. When a pipe's diameter exceeds a certain size, usually either 8 or 10 inches, manholes replace cleanouts. Manholes are typically required every 300 to 400 feet. Check your local code requirements. In addition, they are required at all changes in direction, elevation, grade, and size. Manholes must be protected against flooding and equipped with covers to prevent the escape of gases. Connections with manholes are often required to be made with flexible compression joints. These connections must not be closer than 1 foot to the manhole and not further than 3 feet away.

TRAPS

Traps are required on drainage-type plumbing fixtures. No fixture is allowed to be double-trapped, and traps serving automatic clothes washers or laundry tubs must not discharge into a kitchen sink. With some fixtures, such as toilets, traps are not apparent because they are an integral component. The following regulations do not apply to integral traps, which are governed by regulations controlling the use of approved fixtures. Drawn-brass tubing traps are not allowed for use with urinals.

Every trap for every fixture is required to have a trap seal that is made with a liquid, usually water, that is not less than 2 inches and not more than 4 inches in depth. In special cases, the depth of a trap seal may vary. If there is a possibility that a trap seal will be compromised through evaporation, the trap must be equipped with a primer that will maintain the trap seal.

Fast Fact

Every trap for every fixture is required to have a trap seal that is made with a liquid, usually water, that is not less than 2 inches and not more than 4 inches in depth. In special cases, the depth of a trap seal may vary.

Trade Tip

P-traps must be properly vented. Without adequate venting, the trap seal can be removed by backpressure.

P-Traps

P-traps are the traps most frequently used in modern plumbing systems. These traps are self-cleaning and frequently have removable U-bends that may act as cleanouts, pending local approval. Fixture traps must be self-scouring and are not allowed to have interior partitions. An exception concerning interior partitions comes into play with integral traps and traps that are constructed of an approved material that is resistant to corrosion and degradation. P-traps must be properly vented. Without adequate venting, the trap seal can be removed by backpressure. Slip joints must be made with an approved elastomeric gasket and can only be installed on the trap inlet, trap outlet, and within the trap seal.

S-Traps

S-traps were very common when most plumbing drains came up through the floor instead of out from a wall. Many S-traps are still in operation, but they are no longer allowed in new installations. S-traps are subject to losing their trap seal through self-siphoning.

Drum Traps

Drum traps are not normally allowed in new installations without special permission from the code officer. The only occasion when drum traps are still used frequently is when they are installed with a combination waste and vent system. It is acceptable to use drum traps as solids interceptors and when they serve chemical-waste systems.

Bell Traps

Bell traps are not allowed for use in new installations.

House Traps

House traps are no longer allowed; they represent a double trapping of all fixtures. Local codes may allow house traps under certain circumstances. House traps were once installed where the building drain joined with the sewer. Most house traps were installed inside the structure, but a fair number were installed outside underground. Their purpose was to prevent sewer gas from coming out of the sewer and into the plumbing system. But house traps make drain cleaning very difficult, and they create a double-trapping situation, which is not allowed. This regulation, like most regulations, is subject to amendment and variance by the local code official.

Crown-Vented Traps

Crown-vented traps are not allowed in new installations. These traps have a vent rising from the top of the trap. As you learned earlier, crown venting must be done at the trap arm, not the trap.

Other Traps

Traps that depend on moving parts or interior partitions are not allowed in new installations.

Individual Traps

Basically, every fixture requires an individual trap, but there are exceptions. One such exception is the use of a continuous waste to connect the drains from multiple sink bowls to a common trap. This is done frequently with kitchen sinks.

There are some restrictions involving the use of continuous wastes. Let's take a kitchen sink as an example. When you have a double-bowl sink, it is okay to use a continuous waste as long as the drains from each bowl are no more than 30 inches apart and neither bowl is more than 6 inches deeper than the other bowl. Some jurisdictions require that all sinks connected to a continuous waste must be of equal depth. Exceptions to this rule do exist.

What if your sink has three bowls? Three-compartment sinks may be connected with a continuous waste. You may use a single trap to collect the drainage from up to three separate sinks or lavatories, as long as the sinks

Fast Fact

Traps must be installed level in order for the trap seal to function properly.

or lavatories are next to each other and in the same room. But the trap must be in a location central to all sinks or lavatories.

Trap Sizes

Trap sizes are determined by the local code. A trap may not be larger than the drainpipe it discharges into.

Tailpiece Length

The tailpiece between a fixture drain and the fixture's trap may not exceed 24 inches.

Standpipe Height

A standpipe, when installed, must extend at least 18 inches but not more than 42 inches above the trap. The standpipe should not extend more than 4 feet from the trap. Some local codes require that a standpipe not exceed a height of 2 feet above the trap. Plumbers installing laundry standpipes often forget this regulation. When setting your fitting height in the drainage pipe, keep in mind the height limitations on your standpipe. Otherwise, your takeoff fitting may be too low or too high to allow your standpipe receptor to be placed at the desired height. Traps for kitchen sinks may not receive the discharge from a laundry tub or clothes washer.

Installation

There is more to proper trap installation than location and trap selection. Traps must be installed level in order for the trap seal to function properly. An average trap seal will consist of 2 inches of water. Some large traps may have a seal of 4 inches, and where evaporation is a problem, deep-sealing traps may have a deeper water seal. The positioning of the trap is critical for the proper seal. If the trap is cocked, the water seal will not be uniform and may contribute to self-siphoning.

When a trap is installed below grade and must be connected from above grade, the trap must be housed in a box of some kind. An example of such a situation would be a trap for tub waste. When installing a bathtub on a concrete floor, the trap is located below the floor. Since the trap cannot be reasonably installed until after the floor is poured, access must be made for the connection. This access, frequently called a tub box or trap box, must provide protection against water, insect, and rodent infiltration.

GREASE TRAPS

One type of trap we have not yet discussed is a grease trap. The reason we haven't talked about grease traps is that they are not really traps; they are interceptors. They are frequently called grease traps, but they are actually grease interceptors. There is a big difference between a trap and an interceptor. Grease traps must conform to PDI G101 and must be installed in accordance with the manufacturer's instructions.

Grease traps must be equipped with devices to control the rate of water flow so that it does not exceed the rated flow of the trap. A flow-control device must be vented. The vent cannot terminate less than 6 inches above the flood-rim level and must be installed in accordance with the manufacturer's instructions.

The vented flow-control device must be located so that there are no system vents between the flow control and the grease-trap inlet. The vent or air inlet of the flow-control device must connect with the sanitary drainage vent system as elsewhere required by the code or terminate through the roof of the building and not to the free atmosphere inside the building.

Trade Tip

Grease traps must be equipped with devices to control the rate of water flow so that the water flow does not exceed the rated flow of the trap.

Traps are meant to prevent sewer gas from entering a building. Traps do not restrict what goes down the drain, only what comes up the drain. Of course, traps do prevent objects larger than the trap from entering the drain, but this is not their primary objective.

Interceptors, on the other hand, are designed to control what goes down a drain. Interceptors are used to keep harmful substances from entering the sanitary drainage system. Separators, because they separate the materials entering them and retain certain materials while allowing others to continue into the drainage system, are also required in some circumstances. Interceptors are used to control grease, sand, oil, and other materials.

Interceptors and separators are required when conditions provide an opportunity for harmful or unwanted materials to enter a sanitary drainage system. When oil, grease, sand, or other harmful substances are likely to enter a drainage system, an interceptor is required. For example, a restaurant is required to be equipped with a grease interceptor. An oil separator would be required for a building where automotive repairs are made. Interceptors and separators must be designed for each individual situation. There is no rule-of-thumb method for choosing the proper interceptor or separator without expert design.

There are some guidelines provided in plumbing codes for interceptors and separators. The capacity of a grease interceptor is based on two factors, grease retention and flow rate. Capacity determinations are typically made by a professional designer. The size of a receptor or separator is also normally determined by a design expert.

CODE UPDATE

The vertical distance from a fixture outlet to the trap weir must not be more than 24 inches. The horizontal distance cannot exceed 30 inches when measured from the centerline of a fixture outlet to the centerline of the inlet of the trap. Standpipes for washing machines are not subject to this rule.

A grease trap or grease interceptor is required to receive the drainage from fixtures and equipment with grease-laden waste located in food-preparation areas, such as in restaurants, hotel kitchens, hospitals, and so forth.

Where food-waste grinders connect to grease traps or grease interceptors, the interceptors must be sized and rated for the discharge of the food-waste grinder. Grease traps and interceptors are not required in private living quarters and individual dwelling units.

Interceptors for sand and other heavy solids must be readily accessible for cleaning. These units must contain a water seal of not less than 6 inches. Some codes require a minimum water depth of only 2 inches. When an interceptor is used in a laundry, a water seal is not required. Laundry receptors, used to catch lint, string, and other objects, are usually made of wire, and they must be easily removed for cleaning. Their purpose is to prevent solids with a diameter of 0.5 inch or more from entering the drainage system.

Oil separators are required at repair garages; gasoline stations with grease racks, grease pits, or work racks; car-washing facilities with engine or undercarriage cleaning capability; and factories where oily and flammable liquid wastes are produced. The separators must keep oil-bearing, grease-bearing, and flammable wastes from entering the building drainage system or other point of disposal.

Other types of separators are used for various plants, factories, and processing sites. The purpose of all separators is to keep unwanted objects and substances from entering the drainage system. Vents are required if it is suspected that these devices will be subject to the loss of a trap seal. All interceptors and separators must be readily accessible for cleaning, maintenance, and repairs.

BACKWATER VALVES

Backwater valves are essentially check valves. They are installed in drains and sewers to prevent the backing up of waste and water in the drain or sewer. Backwater valves are required to be readily accessible and must be installed whenever a drainage system is likely to encounter backups from the sewer.

The intent behind backwater valves is to prevent sewers from backing up into individual drainage systems. Buildings that have plumbing fixtures below the level of the street where a main sewer is installed are candidates for backwater valves.

This concludes our section on traps, cleanouts, interceptors, and other drainage-related regulations. While this is a short chapter, it is an important one. You may not have a need for installing manholes or backwater valves every day, but, as a plumber, you will frequently work with traps and cleanouts.

CHAPTER

11

Storm Drainage

Storm-water drainage piping is designed to convey excess groundwater to a suitable location. A suitable location might be a catch basin, storm sewer, or pond. Storm-water drainage may never be piped into a sanitary sewer or plumbing system.

When you wish to size a storm-water drainage system, you must have some benchmark information to work with. One consideration is the pitch of a horizontal pipe. Another piece of the puzzle is the number of square feet of surface area your system will be required to drain. You will also need data on the rainfall rates in your area.

When you use your code book to size a storm-water system, you should have access to all the key elements required to size the job except possibly the local rainfall amounts. You should be able to obtain rainfall figures from your state or county offices. Your code book should provide you with a table to use in making your sizing calculations.

SIZING

The first step to take when sizing a storm drain or sewer is to establish your known criteria. How much pitch will your pipe have? Your code book should offer choices for pipe pitch.

What else do you need to know? You must know what the rainfall rate is for the area where you will be installing the storm-water system. There should be a table in your code book that lists many regions and their rates of rainfall. You must also know the surface area that your system will be responsible for handling. The surface area must include both roof and parking areas.

When you are working with a standard table like the ones found in most code books, you must convert the information to suit your local conditions. For example, if a standardized table is based on 1 inch of rainfall an hour and your location has 2.4 inches of rainfall per hour, you must convert the table, but this is not difficult.

When I want to convert a table based on a 1-inch rainfall to meet my local needs, all I have to do is divide the drainage area in the table by my rainfall amount. For example, if my standard chart shows an area of 10,000 square feet requiring a 4-inch pipe, I can change the table by dividing my rainfall amount, 2.4, into the surface area of 10,000 square feet.

If I divide 10,000 by 2.4, I get 4167. All of a sudden, I have solved the mystery of computing storm-water piping needs. With this simple conversion, I know that if my surface area was 4,167 square feet, I would need a 4-inch pipe. But, my surface are is 15,000 square feet, so what size pipe do I need? Well, I know it will have to be larger than 4 inches. So, I look down my conversion chart and find the appropriate surface area. My 15,000 square feet of surface area will require a storm-water drain with a diameter of 8 inches. I found this by dividing the surface areas of the numbers in the table found

CODE UPDATE

Storm drainage systems must be provided with backwater valves.

in my code book by 2.4 until I reached a number equal to or greater than my surface area. I could almost get by with a 6-inch pipe, but not quite.

Now, let's recap this exercise. To size a horizontal storm drain or sewer, decide the pitch you will put on the pipe. Next, determine your area's maximum rainfall for a 1-hour storm (the highest rainfall over the last 100 years). If you live in a city, your city may be listed, with its rainfall amount, in your code book. Using a standardized chart rated for 1 inch of rainfall per hour, divide the surface area by a factor equal to your rainfall index; in my case it was 2.4. This division process converts a generic table into a customized table just for your area.

Once the math is done, look down the table for the surface area that most closely matches the area you have to drain. To be safe, go with a number slightly higher than your projected number. It is better to have a pipe one size too large than one size too small. When you have found the appropriate surface area, look across the table to see what size pipe you need. See how easy that was? Well, maybe it's not easy, but it is a chore you can handle.

Rain Leaders

When you are required to size rain leaders or downspouts, you use the same procedure described above with one exception. You use a table, supplied in your code book, to size the vertical piping. Determine the amount of surface area your leader will drain and use the appropriate table to establish your pipe size. The conversion factors are the same.

Sizing gutters is essentially the same as sizing horizontal storm drains. You will use a different table, provided in your code book, but the mechanics are the same.

ROOF DRAINS

Roof drains are often the starting point of a storm-water drainage system. As the name implies, roof drains are located on roofs. On most roofs, the drains are equipped with strainers that protrude upward at least 4 inches to catch leaves and other debris. Roof drains should be at least twice the size of the piping connected to them. All roofs that do not drain to hanging gutters are required to have roof drains. A minimum of two roof drains should be installed on roofs with a surface area of 10,000 square feet or less.

If the surface area exceeds 10,000 square feet, a minimum of four roof drains should be installed.

When a roof is used for purposes in addition to shelter, the roof drains may have a strainer that is flush with the roof's surface. Roof drains should obviously be sealed to prevent water from leaking around them. The size of the roof drain can be instrumental in the flow rates designed into a storm-water system. When a controlled flow from roof drains is needed, the roof structure must be designed to accommodate the degree of flow.

Secondary roof systems must be equipped with an end point of discharge that is separate from the primary system. The discharge must occur above grade.

Combination Storm Drain and Sewer

If a combined storm-drain and sewer arrangement is approved, it must be sized properly. This requires converting fixture-unit loads into drainage surface area. For example, 256 fixture units will be treated as 1000 square feet of surface area. Each additional fixture unit in excess of 256 will be assigned a value of 3.9 square feet. In the case of sizing for continuous flow, each gpm is rated as 96 square feet of drainage area.

STORM-WATER PIPING

Storm-water piping requires the same number of cleanouts, with the same frequency, as a sanitary system. Just as regular plumbing pipes must be protected, so must storm-water piping. For example, if a downspout is in danger of being crushed by automobiles, you must install a guard to protect it. Backwater valves installed in a storm-drainage system must conform to local code requirements.

Fast Fact

Storm-water piping requires the same number of cleanouts, with the same frequency, as a sanitary system.

Trade Tip

All sump-pump discharge pipes should be equipped with a check valve.

As I said earlier, storm-water and sanitary systems should not be combined. There may be some locations where the two are combined, but they are the exception rather than the rule. Area-way drains or floor drains must be trapped. When rain leaders and storm drains are allowed to connect to a sanitary sewer, they are required to be trapped. The trap must be equal in size to the drain it serves. Traps must be accessible for cleaning the drainage piping. Storm-water piping may not be used for conveying sanitary drainage.

SUMP PUMPS

Sump pumps are used to remove water collected in building subdrains. These pumps must be placed in a sump, but the sump need not be covered with a gas-tight lid or vented. The lid must be removable. Sump pits must be at least 18 inches in diameter and at least 24 inches in depth. Pits must be accessible and installed so that all water entering the pit flows in naturally by gravity. Construction of a sump pit may be accomplished with tile, steel, plastic, cast iron, or concrete.

Many people are not sure what to do with the water pumped out of their basement by a sump pump. Do you pump it into your sewer? No, the discharge from a sump pump should not be pumped into a sanitary sewer. The water from the pump should be pumped to a storm-water drain, or in some cases, to a point on the property where it will not cause a problem.

All sump-pump discharge pipes should be equipped with a check valve. The check valve prevents previously pumped water from running down the discharge pipe and refilling the sump, forcing the pump to pull double duty. When I speak of sump pumps, I am talking about pumps removing groundwater, not waste or sewage.

VARIATIONS

There are some variations in local codes for storm-water drainage. For example, approved materials can differ from one jurisdiction to another. This can be true of both aboveground and underground materials. Once storm-water piping extends at least 2 feet from a building, any approved material may be used in most regions.

Another example of a variation is that the inlet area of a roof drain is generally only required to be one and one-half times the size of the piping connected to the roof drain. However, when positioned on roofs used for purposes other than weather protection, roof-drain openings must be sized twice as large at the drain connecting to them.

Some regions provide different tables for sizing purposes. When computing the drainage area, you must take into account the effect vertical walls have on the drainage area. For example, a vertical wall that reflects water onto the drainage area must be allowed for in your surface-area computations. In the case of a single vertical wall, add one-half of the wall's total square footage to the surface area.

Two vertical walls that are adjacent to each other require you to add 35 percent of the combined wall square footage to your surface area.

If you have two walls of the same height that are opposite each other, no added space is needed. In this case, each wall protects the other and does not allow extra water to collect on the roof area.

When you have two opposing walls with different heights, you must make a surface-area adjustment. Take the square footage of the higher wall above the other wall, and add half the square footage to your surface area.

Fast Fact

Some roof designs require a backup drainage system in case of emergencies. These roofs are generally surrounded by vertical sections.

When you encounter three walls, you use a combination of the above instructions to reach your goal. Four walls of equal height do not require an adjustment. If the walls are not of equal height, use the procedures listed above to compute your surface area.

Additional code variations may occur with sump pits. Some sump pits are required to have a minimum diameter of 18 inches. In some regions, floor drains may not connect to drains intended solely for storm water. When computing surface area to be drained for vertical walls, such as walls enclosing a rooftop stairway, use one-half the total square footage from the vertical wall surface that reflects water onto the drainage surface.

Some roof designs require a backup drainage system in case of emergencies. These roofs are generally surrounded by vertical sections. If these vertical sections are capable of retaining water on the roof if the primary drainage system fails, a secondary drainage system is required. In these cases, the secondary system must have independent piping and discharge locations. These special systems are sized by using different rainfall rates. The ratings are based on a 15-minute rainfall. Otherwise, the hundred-year conditions still apply.

Some regions have requirements for sizing a continuous flow that provide a rating of 24 square feet of surface area for every gpm generated. For regular sizing based on 4 inches of rain per hour, 256 fixture units equal 1000 square feet of surface area. Each additional fixture unit is rated at 3.9 inches. If the rainfall rate varies, a conversion must be done.

To convert the fixture-unit ratings to a higher or lower rainfall, you must do some math. Take the square-foot rating assigned to fixture units and multiply it by four. For example, 256 fixture units equal 1000 square feet. Multiply 1000 by four to get 4000. Now divide 4000 by the rate of rainfall for one hour. Say, for example, that the hourly rainfall was 2 inches; the converted surface area would be 2000.

You have made it past a section of code regulations that gives professional plumbers the most trouble. Storm-water drains are despised by some plumbers, because they have little knowledge of how to compute them. With the aid of this chapter, you should be able to design a suitable system with minimal effort.

CHAPTER

12

Special Piping and Storage Systems

Medical gas systems and nonmedical oxygen systems are covered in the plumbing code under the provisions for special piping and storage systems or under the category of healthcare facilities and medical gas and vacuum systems, depending upon which local code you are working with. There is a distinction between medical gases and oxygen systems that are not used for medical purposes.

The general provisions of special piping and storage systems govern the design and installation for nonflammable medical gas systems and nonmedical oxygen systems. It's important that you note the limitations of this part of the code. Pay attention to the part about nonflammable medical gas systems and nonmedical oxygen systems. These two elements are all that are covered under the special piping and storage systems as discussed in the plumbing code.

Fast Fact

Drinking fountains, valves, and other items the might normally protrude from a wall must be flush-mounted or full-recessed in corridors and other areas where patients may be transported on a gurney, hospital bed, or wheelchair.

REQUIREMENTS

Drinking fountains, valves, and other items the might normally protrude from a wall must be flush-mounted or full-recessed in corridors and other areas where patients may be transported on a gurney, hospital bed, or wheelchair. Piping and traps in psychiatric patient rooms must be concealed. All fixtures and fittings in these rooms must be vandal-proof. All ice makers or ice-storage chests must be installed in a nurse station or other similarly supervised area that is not subject to contamination.

Sterilizers

Drains from sterilizers must be connected to a drainage system through an indirect-waste connection. The indirect-waste pipe must not be less than the size of the drain connection on the fixture being served. The length of the piping must not exceed 15 feet in length. Receptors must be located in the same room as the equipment served. With the exception of bedpan steamers, such indirect-waste pipes do not require traps. A trap with a minimum seal of 3 inches must be provided in the indirect-waste pipe for a bedpan steamer. Sterilizers that have provisions for a vapor vent required by the manufacturer must be extended to the outdoors above the roof. These vents are not allowed to be connected to any drainage-system vent.

Fast Fact

Drains from sterilizers must be connected to a drainage system through an indirect-waste connection.

Trade Tip

Aspirators used for removing body fluids must be equipped with a collecting bottle or similar fluid trap.

Aspirators

All aspirators or other water-supplied suction devices can only be installed in strict accordance with code requirements. Aspirators used for removing body fluids must be equipped with a collecting bottle or similar fluid trap. Aspirators must indirectly discharge to the sanitary drainage system through an air gap. The potable-water supply to an aspirator must be protected by a vacuum breaker or equivalent backflow protection.

MEDICAL GASES

Medical gases covered under this portion of the code are nonflammable. Key components of the code pertain to nonflammable medical gas systems generally, inhalation anesthetic systems, and vacuum piping systems. These systems might exist in hospitals, dental offices, or other facilities. The code rulings are simple. These systems must be designed and installed in accordance with NFPA 99C. There are, however, two exceptions. The special piping and storage-systems section of the code does not apply to portable systems or cylinder storage. Vacuum-system exhaust must comply with the local mechanical code.

OXYGEN SYSTEMS

Oxygen systems that are not used for medical purposes must be designed and installed in accordance with NFPA 50 and NFPA 51. It may strike you as strange, but this is all that the code has to say about nonmedical oxygen systems.

MORE DETAILED REQUIREMENTS

Some codes have much more detailed code requirements for healthcare installations, and we are going to review them now. As you would expect, all medical gas and medical vacuum systems must be installed in accordance with all code requirements. The code requires all installers to be competent. Medical gas and vacuum systems must be supplied with at least two sources. For example, a system would be required to have at least two cylinder banks with at least two cylinders in each bank, a minimum of two air compressors, or a minimum of two vacuum pumps. However, two supply pipelines are not required. Operating pressures, minimum flow rates, and minimum station outlets and inlets are regulated by code requirements.

The sizing of medical gas and vacuum systems should be done by a mechanical engineer. This recommendation appears in the code book. Since the code suggests that only mechanical engineers be responsible for the design of the systems, I see no reason to cover sizing methods in this book. The local code book does offer information on sizing requirements and practices, but, again, the code clearly recommends that the sizing by done by an engineer.

Plans and specs must be provided to the local code officer prior to installing any medical gas or medical vacuum system. The plans and specs must be approved prior to the issuance of a permit. An approval package will normally contain a plot plan of the site, drawn to scale, that indicates the location of existing or new cylinder storage areas, property lines, driveways, and existing or proposed buildings. There will be a piping layout of all proposed piping systems and alterations. Full specifications of the materials to be used are required. A record of as-built plans and valve-identification records must remain on the site of the system at all times. Always check your local code requirements prior to performing any work. Codes vary, and you must check your local requirements to be sure of your obligations.

CHAPTER

13

Recycling Gray Water

The recycling of gray water has become more popular, and the plumbing code has responded to this demand. What is gray water? It is defined as waste water that is discharged from lavatories, bathtubs, showers, clothes washers, and laundry sinks. Gray water is not a new issue. Dry wells have been used to collect gray water for years. However, the modern plumbing code is much more restrictive in the recycling of gray water. This is done for public health and safety and should not be considered a bad position.

The plumbing code generally states that all plumbing fixtures must discharge into a sanitary drainage system. There is an exception that allows for the recycling of gray water. Fixtures that are eligible for gray water recycling are as follows:

- Bathtubs
- Showers
- Lavatories
- Clothes washers
- Laundry sinks.

CODE UPDATE

Collection reservoirs must be equipped with overflow pipes that have the same, or larger, diameters as the influent pipe for grey water. The overflow pipe is to be connected indirectly to a sanitary drainage system.

What can gray water be used for when it is recycled? The water can be used to flush water closets and urinals that are located in the same building as the gray-water recycling system. All of these systems must comply with the overall regulations of the plumbing code. In addition to being used for flushing toilets and urinals, gray water can be used for irrigation purposes when specific approval is given by the administrating authority.

Any installation involving the use of gray water, including drains, wastes, and vent pipes, must be installed in full compliance with the plumbing code. Gray water is to be collected in a reservoir. The reservoir must be constructed of durable, nonabsorbent, and corrosion-resistant materials. It is required that the reservoir be a closed and gas-tight vessel.

Access openings are required to allow inspection and cleaning of the reservoir interior. Each reservoir needs a holding capacity that is at a minimum twice the size of the volume of water required to meet the daily flushing requirements of the fixtures supplied with gray water but not less than 50 gallons. Each reservoir must be sized to limit retention time of gray water to a maximum of 72 hours.

All gray water that enters a reservoir must pass through an approved filter such as a media, sand, or diatomaceous-earth filter. A disinfection process is required for all gray water that is recycled. The disinfection process must be of an approved type. Three approved types of disinfectants include chlorine, iodine, and ozone.

It' s not always possible to collect and recycle enough gray water for the intended purposes. This requires the use of what is known as makeup water. Potable water is added to gray water in order to create the makeup factor.

When this is the case, the potable-water supply must be protected against backflow in accordance with code requirements. Any potable-water supply that feeds a gray-water reservoir must be fitted with a full-open valve.

An overflow pipe must be installed in conjunction with a reservoir collection point. The overflow pipe must be of the same diameter s the influent pipe for the gray water. Any overflow pipe must be connected directly to a sanitary drainage system.

All collection reservoirs must be fitted with a drain at the lowest point of the reservoir. The drain must be connected directly to a sanitary drainage system. It is required that the drain be the same size as the overflow pipe required, and the drain must be equipped with a full-open valve.

A vent is required for a collection reservoir. The sizing of the vent is determined by normal code requirements, based on the size of the reservoir influent pipe. All gray water used in a recycling situation is required to be dyed either blue or green. The dyeing agent must be a food-grade vegetable dye. Gray water must be dyed prior to the water being supplied to plumbing fixtures.

Piping that is used for the distribution of gray water and reservoirs must be identified as conduits of nonpotable water. Requirements for marking and identifying said piping are covered in the general plumbing code.

Most plumbers don't encounter many gray-water recycling systems. You may never deal with one, but if you do, you must be able to understand the code requirements associated with such a system. Remember that you cannot treat a gray-water system exactly as you would a traditional system. The rules pertaining to gray water are not complex, so don't be intimidated by them. It's all a matter of understanding code requirements, and the regulations for gray water are fairly easy to comprehend.

CODE UPDATE

Gray water must be disinfected by an approved method. Acceptable disinfectants include chlorine, iodine, and ozone.

CHAPTER

14

Using the Code In the Real World

Using the code in the real world is a little different from reading and understanding a code book. In theory, the book should hold all the answers, and you should be able to apply them without incurring any unexpected problems. But reality is not always so simple. Putting the plumbing code to practical use doesn't have to be difficult, but it can be.

There are two extremes to applying code requirements on a job. On one side, you have situations where the code enforcement is lax. Then there are times when the code enforcement is extremely strict. Most jobs run somewhere between the two extremes. Whether the job you are working on is lax or strict, you could have problems dealing with code requirements.

LAX JOBS

There are some plumbers who welcome lax jobs. These plumbers enjoy not having their work scrutinized too tightly. I guess everyone might enjoy an easy job, but there are risks to lax jobs.

In my opinion professional plumbers should perform professional services, even if they can get away with less-than-credible work. This should be an ethical commitment. But there are plumbers who will cut corners in jurisdictions where they know that they can get away with it. This is not fair to the customer, and it can put the plumber at risk.

Inspectors provide a form of protection for plumbers and plumbing contractors. Cutting corners can put you at great risk for a lawsuit. Leaving the financial side of such a suit out of the picture, try to imagine how you would deal with the guilt of personal injury to people as a result of your deviation from the plumbing code.

How many times have you installed a replacement water heater without a permit? You know that a permit and inspection are required. Many plumbers have skipped the permit and inspection element of this type of job at one time or another. But they probably never thought of the risk. Let me give you an example of how such a code violation could cost you more than you might ever imagine.

Let's say that you install a water heater without the required permit and inspection. We will say that you do this to keep the cost down for your customer. It's a simple, standard replacement job. What could go wrong? You do the job outside the code requirements for a permit and inspection. You test the installation and all is well. The customer is satisfied and you leave.

Several months after your installation, there is a significant problem with the water heater that you installed against code requirements. The problem could be a ruptured pipe that floods the home, an electrical fire, or worse. What do you think your insurance carrier is going to say when they find out that you did an illegal installation? It won't be what you want to hear.

Your liability insurance may not be willing to cover any claim made for work that you did without following code requirements. If you had done the job by the book, you would have some foundation for defending yourself. Additionally, you would have an approved inspection certificate that would validate the fact that you did the work properly and within code requirements. This could go a long way in a lawsuit. The small amount of money you save by cheating could come back as a massive lawsuit that could wipe out your career or your business.

When you work in an area prone to lax inspections, you can be at risk. It is in your best interest to do all work to code requirements and to have that work inspected and approved. How much noise are you going to make if this situation is not feasible? This is an interesting question that can present you with other problems. I will give you two examples to consider.

I was working as a consultant with a plumbing company several years ago. The company was doing remodeling and new installations in a restaurant. When the company called for a final inspection, the inspector was lazy. He asked if everything was installed and working properly. It was. The inspector said he would mail the approval slip for the final inspection to the company. Would you let a job like this close without a site inspection by a code-enforcement officer? It's not good. But are you going to complain and get the inspectors in the area against you? This is the difficult part. You have to make this judgment call. Personally, I would require a real inspection.

In another case, I responded to the call of a homeowner who was experiencing a bad smell in his new home. I got to the house and found an almost unbelievable situation. The house was new. It was built on a pier foundation. When I pulled into the driveway, there were two small children splashing around in what appeared to be a puddle. I met with the homeowner and then went to investigate the problem.

I walked around the home, and the problem was obvious. Some idiot had run slotted pipe from the toilet to the septic tank. The splashing children were dancing in sewage! To make matters worse, there were wetlands within a few dozen feet of the open sewage. When I returned to my office, I called the state plumbing inspector and reported my finding. The response astounded me. I was told that there was nothing that could be done about the problem since it was on private property. This just is not right.

STRICT CODE ENFORCEMENT

Strict code enforcement can be a blessing. It helps to take you off the hook if something goes wrong down the road. On the other hand, it can be a real pain when you are being ripped apart on a simple job. Again, you have to use common sense in your response to such situations.

Should an inspector fail your job because you are missing one nail plate? Technically, yes. In reality, especially if the code officer knows that you are reputable and professional, what would it hurt to pass the job with a notice for you to install the missing nail plate?

Do plumbing inspectors really have to put a grade level on all the drainage pipe that you install? Probably not. Can they? Sure they can. Where is the line drawn? I believe the line is drawn when inspectors abuse their power, but this is just my personal opinion. Let me give you an example.

I installed an island sink many years ago. The code book at that time had a diagram of exactly how the piping system should be configured. As a young master plumber, I followed the diagram to the letter. The work looked like the drawing in the code book and I was quite pleased. Then the plumbing inspector arrived and my feelings changed abruptly.

When the plumbing inspector arrived, he looked under the sink and saw the piping arrangement. He turned to me and indicated that the work looked like spaghetti. I was shocked. The inspector started writing a rejection slip for the job. In the meantime, I went to my truck and got my code book to confront the code officer.

As the inspector came out of the house, I approached him with my code book and tried to show him the diagram. He would not even look at the book. I was very angry and drove directly to the code-enforcement office. When I arrived, I went to the supervisor's office with my code book and explained the situation.

The supervisor reached the inspector in the field and had him meet us back at the job. The supervisor looked at the work and said it was fine. He approved the job. I felt great. The first inspector was not happy. It seemed like a victory for me, but I paid for it over the next year or so. I never experienced so many picky rejections in my life as I did after that situation.

Should I have kept my mouth shut? Maybe, but I don't stay quiet well when I know I am right. Still, my proving that I was right was a costly mistake with regard to future jobs. This type of situation has to be weighed when making business decisions.

SAFETY

The plumbing code exists for a reason. Safety is what the code is focused on, and this is a valid reason for code regulations. Most plumbers have felt from one time to another that code regulations are too strict or not needed for certain elements of the business. While this may be true on occasion, the basic foundation of the code is a sound one, and it should be observed. There are elements of code regulations that I feel are too detailed for practical purposes, but I always attempt to comply with code requirements.

FEES

The fees charged for permits are often complained about by both contractors and homeowners. These fees pay for the administration of the code and the protection of consumers. Indirectly, contractors receive protection from inspections. When you are feeling taken advantage of due to costly fees charged by a code-enforcement office, consider the benefits that our society is receiving. I am not trying to be a proponent of code fees, but I do believe that people in the trades should look at the fees in a reasonable fashion.

KNOW YOUR INSPECTORS

To make your life easier, get to know your inspectors. Communication is a major portion of any good deal. Knowing the inspectors in your area and what they expect will make jobs run more smoothly. Most inspectors are accessible and helpful as long as you don't approach them with a bad attitude.

LOCAL JURISDICTIONS

Local jurisdictions adopt a code and have the right to amend it for local requirements. In simple terms, this can mean that a local code may vary from the base code that is adopted. For example, a vent terminal in some regions may be required to extend 12 inches above a roof. In other regions, the same type of vent may have to extend 24 inches above a roof. The reason is that heavy snow loads may cover shorter vents. It is a regional change. Even if you know the primary code by heart, you might have a conflict with a local jurisdiction. A visit with your local code officers will clear up such differences before they become a problem on a job.

COMMON SENSE

Common sense goes a lot way in the installation of plumbing systems. It is not uncommon for a mixture of common sense and code requirements to be used to achieve a successful installation. Code regulations must be followed, but they are not always as rigid as they may appear to be. Don't be afraid to consult with inspectors to arrive at a reasonable solution to difficult problems.

In closing, using the code in the real world is largely a matter of proper communication between plumbing contractors and code officers. This is an important fact to keep in mind. Learn the code and use it properly. If you feel a need to deviate from it, talk with a code officer and arrive at a solution that is acceptable to all parties affected. If you keep a good attitude and don't scrap the code because you think you know better, you should do well in your trade.

APPENDIX

Referenced Standards

ANSI

American National Standards Institute
25 West 43rd Street, Fourth Floor
New York, NY 10036

Standard Reference Number	Title	Referenced in code section number
A118—10.99	Specifications for Load Bearing, Bonded, Waterproof Membranes for Thin Set Ceramic Tile and Dimension Stone Installation.	417.5.2.5
Z4.3—95	Minimum Requirements for Nonsewered Waste-disposal Systems	311.1
Z21.22—99 (R2003)	Relief Valves for Hot Water Supply Systems with Addenda Z21.22a-2000 (R2003) and Z21.22b-2001 (R2003).	504.2, 504.4
Z124.1—95	Plastic Bathtub Units	407.1
Z124.2—95	Plastic Shower Receptors and Shower Stalls	417.1
Z124.3—95	Plastic Lavatories	416.1, 416.2, 417.1
Z124.4—96	Plastic Water Closet Bowls and Tanks	420.1
Z124.6—97	Plastic Sinks	415.1, 418.1
Z124.9—94	Plastic Urinal Fixtures	419.1

AHRI

Air-Conditioning, Heating, & Refrigeration Institute
4100 North Fairfax Drive, Suite 200
Arlington, VA 22203

Standard Reference Number	Title	Referenced in code section number
1010—02	Self-contained, Mechanically Refrigerated Drinking-Water Coolers	410.1

ASME

American Society of Mechanical Engineers
Three Park Avenue
New York, NY 10016-5990

Standard Reference Number	Title	Referenced in code section number
A112.1.2—2004	Air Gaps in Plumbing Systems	Table 608.1, 608.13.1
A112.1.3—2000 Reaffirmed 2005	Air Gap Fittings for Use with Plumbing Fixtures, Appliances and Appurtenances	Table 608.1, 608.13.1
A112.3.1—2007	Stainless Steel Drainage Systems for Sanitary, DWV, Storm and Vacuum Applications Above and Below Ground	412.1, Table 702.1, Table 702.2, Table 702.3, Table 702.4, 708.2, Table 1102.4, Table 1102.5, 1102.6, Table 1102.7
A112.3.4—2000 (Reaffirmed 2004)	Macerating Toilet Systems and Related Components	712.4.1
A112.4.1—1993 (R2002)	Water Heater Relief Valve Drain Tubes	504.6
A112.4.3—1999 (Reaffirmed 2004)	Plastic Fittings for Connecting Water Closets to the Sanitary Drainage System	405.4
A112.6.1M—1997 (R2002)	Floor-affixed Supports for Off-the-floor Plumbing Fixtures for Public Use	405.4.3
A112.6.2—2000 (Reaffirmed 2004)	Framing-affixed Supports for Off-the-floor Water Closets with Concealed Tanks	405.4.3
A112.6.3—(Reaffirmed 2007)	2001 Floor and Trench Drains	412.1
A112.6.7—2001 (Reaffirmed 2007)	Enameled and Epoxy-coated Cast-iron and PVC Plastic Sanitary Floor Sinks	427.1
A112.14.3—2000	Grease Interceptors	1003.3.4
A112.14.4—2001 (Reaffirmed 2007)	Grease Removal Devices	1003.3.4
A112.18.1-2005/ CSA B125.1-2005	Plumbing Supply Fittings	424.1, 424.2, 424.3, 607.4, 608.2
A112.18.2-2005/ CSA B125.2-2005	Plumbing Waste Fittings	424.1.2
A112.18.3—2002	Performance Requirements for Backflow Protection Devices and Systems in Plumbing Fixture Fittings	424.2, 424.6
A112.18.6—2003	Flexible Water Connectors	605.6
A112.18.7—1999 (Reaffirmed 2004)	Deck mounted Bath/Shower Transfer Valves with Integral Backflow Protection	424.8
A112.19.1M—2004 (Reaffirmed 2004)	Enameled Cast Iron Plumbing Fixtures	407.1, 410.1, 415.1, 416.1, 418.1
A112.19.2—2003	Vitreous China Plumbing Fixtures and Hydraulic Requirements for Water Closets and Urinals	401.2, 405.9, 408.1, 410.1, 416.1, 418.1, 419.1, 420.1
A112.19.3M—2000 (Reaffirmed 2007)	Stainless Steel Plumbing Fixtures (Designed for Residential Use)	405.9, 415.1, 416.1, 418.1
A112.19.4M—1994 (Reaffirmed 2004)	Porcelain Enameled Formed Steel Plumbing Fixtures	407.1, 416.1, 418.1
A112.19.5—2005	Trim for Water-closet Bowls, Tanks and Urinals	425.4
A112.19.6—1995	Hydraulic Performance Requirements for Water Closets and Urinals	419.1, 420.1
A112.19.7M—2006	Hydromassage Bathtub Appliances	421.1
A112.19.8M—2007	Suction Fittings for Use in Swimming Pools, Wading Pools, Spas, Hot Tubs	421.4
A112.19.9M—1991(R2002)	Nonvitreous Ceramic Plumbing Fixtures with 2002 Supplement	407.1, 408.1, 410.1, 415.1, 416.1, 417.1, 418.1, 420.1
A112.19.12—2006	Wall Mounted and Pedestal Mounted, Adjustable, Elevating, Tilting and Pivoting Lavatory, Sink and Shampoo Bowl Carrier Systems and Drain Systems	416.4, 418.3
A112.19.13—2001 (Reaffirmed 2007)	Electrohydraulic Water Closets	420.1
A112.19.15— 2005	Bathtub/Whirlpool Bathtubs with Pressure Sealed Doors	407.4, 421.5
A112.19.19—2006	Vitreous China Nonwater Urinals	419.1
A112.21.2M—1983	Roof Drains	1102.6
A112.36.2M—1991(R2002)	Cleanouts	708.2
B1.20.1—1983(R2006)	Pipe Threads, General Purpose (inch)	605.10.3, 605.12.3, 605.14.4, 605.16.3, 605.18.1, 705.2.3, 705.4.3, 705.9.4, 705.12.1, 705.14.3
B16.3—2006	Malleable Iron Threaded Fittings Classes 150 and 300	Table 605.5, Table 702.4, Table 1102.7
B16.4—2006	Gray Iron Threaded Fittings Classes 125 and 250	Table 605.5, Table 702.4, Table 1102.7
B16.9—2003	Factory-made Wrought Steel Buttwelding Fittings	Table 605.5, Table 702.4, Table 1102.7
B16.11—2005	Forged Fittings, Socket-welding and Threaded	Table 605.5, Table 702.4, Table 1102.7
B16.12—1998 (Reaffirmed 2006)	Cast-iron Threaded Drainage Fittings	Table 605.5, Table 702.4, Table 1102.7
B16.15—2006	Cast Bronze Threaded Fittings	Table 605.5, Table 702.4, Table 1102.7
B16.18—2001 (Reaffirmed 2005)	Cast Copper Alloy Solder Joint Pressure Fittings	Table 605.5, Table 702.4, Table 1102.7
B16.22—2001 (Reaffirmed 2005)	Wrought Copper and Copper Alloy Solder Joint Pressure Fittings	Table 605.5, Table 702.4, Table 1102.7
B16.23—2002 (Reaffirmed 2006)	Cast Copper Alloy Solder Joint Drainage Fittings DWV	Table 605.5, Table 702.4, Table 1102.7
B16.26—2006	Cast Copper Alloy Fittings for Flared Copper Tubes	Table 605.5, Table 702.4, Table 1102.7
B16.28—1994	Wrought Steel Buttwelding Short Radius Elbows and Returns	Table 605.5, Table 702.4, Table 1102.7
B16.29—2001	Wrought Copper and Wrought Copper Alloy Solder Joint Drainage Fittings (DWV)	Table 605.5, Table 702.4, Table 1102.7

ASSE

American Society of Sanitary Engineering
901 Canterbury Road, Suite A
Westlake, OH 44145

Standard Reference Number	Title	Referenced in code section number
1001—02	Performance Requirements for Atmospheric Type Vacuum Breakers	425.2, Table 608.1, 608.13.6, 608.16.4.1
1002—99	Performance Requirements for Antisiphon Fill Valves (Ballcocks) for Gravity Water Closet Flush Tanks	425.3.1, Table 608.1
1003—01	Performance Requirements for Water Pressure Reducing Valves	604.8
1004—90	Performance Requirements for Backflow Prevention Requirements for Commercial Dishwashing Machines	409.1
1005—99	Performance Requirements for Water Heater Drain Valves	501.3
1006—89	Performance Requirements for Residential Use Dishwashers	409.1
1007—92	Performance Requirements for Home Laundry Equipment	406.1, 406.2
1008—89	Performance Requirements for Household Food Waste Disposer Units	413.1
1009—90	Performance Requirements for Commercial Food Waste Grinder Units	413.1
1010—04	Performance Requirements for Water Hammer Arresters	604.9
1011—04	Performance Requirements for Hose Connection Vacuum Breakers	Table 608.1, 608.13.6
1012—02	Performance Requirements for Backflow Preventers with Intermediate Atmospheric Vent	Table 608.1, 608.13.3, 608.16.2
1013—05	Performance Requirements for Reduced Pressure Principle Backflow Preventers and Reduced Pressure Fire Protection Principle Backflow Preventers	Table 608.1, 608.13.2, 608.16.2
1015—05	Performance Requirements for Double Check Backflow Prevention Assemblies and Double Check Fire Protection Backflow Prevention Assemblies	Table 608.1, 608.13.7
1016—96	Performance Requirements for Individual Thermostatic, Pressure Balancing and Combination Control Valves for Individual Fixture Fittings	424.3, 424.4, 607.4
1017—03	Performance Requirements for Temperature Actuated Mixing Valves for Hot Water Distribution Systems	501.2, 613.1
1018—01	Performance Requirements for Trap Seal Primer Valves; Potable Water Supplied	1002.4
1019—04	Performance Requirements for Vacuum Breaker Wall Hydrants, Freeze Resistant, Automatic Draining Type	Table 608.1, 608.13.6
1020—04	Performance Requirements for Pressure Vacuum Breaker Assembly	Table 608.1, 608.13.5
1022—03	Performance Requirements for Backflow Preventer for Beverage Dispensing Equipment	Table 608.1, 608.16.1, 608.16.10
1024—04	Performance Requirements for Dual Check Valve Type Backflow Preventers (for Residential Supply Service or Individual Outlets)	605.3.1, Table 608.1
1035—02	Performance Requirements for Laboratory Faucet Backflow Preventers	Table 608.1, 608.13.6
1037—90	Performance Requirements for Pressurized Flushing Devices for Plumbing Fixtures	425.2
1044—01	Performance Requirements for Trap Seal Primer Devices Drainage Types and Electronic Design Types	1002.4
1047—05	Performance Requirements for Reduced Pressure Detector Fire Protection Backflow Prevention Assemblies	Table 608.1, 608.13.2
1048—05	Performance Requirements for Double Check Detector Fire Protection Backflow Prevention Assemblies	Table 608.1, 608.13.7
1050—02	Performance Requirements for Stack Air Admittance Valves for Sanitary Drainage Systems	917.1
1051—02	Performance Requirements for Individual and Branch Type Air Admittance Valves for Sanitary Drainage Systems-fixture and Branch Devices	917.1
1052—04	Performance Requirements for Hose Connection Backflow Preventers	Table 608.1, 608.13.6
1055—97	Performance Requirements for Chemical Dispensing Systems	608.13.9
1056—01	Performance Requirements for Spill Resistant Vacuum Breaker	Table 608.1, 608.13.5, 608.13.8
1060—96	Performance Requirements for Outdoor Enclosures for Backflow Prevention Assemblies	608.14.1
1061—06	Performance Requirements for Removable and Nonremovable Push Fit Fittings	Table 605.5
1062—97	Performance Requirements for Temperature Actuated, Flow Reduction Valves to Individual Fixture Fittings	424.7
1066—97	Performance Requirements for Individual Pressure Balancing In-line Valves for Individual Fixture Fittings	604.11
1069—05	Performance Requirements for Automatic Temperature Control Mixing Valves	424.4
1070—04	Performance Requirements for Water-temperature Limiting Devices	408.3, 416.5, 424.5, 607.1
1072—06	Performance Requirements for Barrier Type Floor Drain Trap Seal Protection Devices	1002.4
1079—2005	Dielectric Pipe Unions	605.24.1, 605.24.3
5013—98	Performance Requirements for Testing Reduced Pressure Principle Backflow Prevention Assembly (RPA) and Reduced Pressure Fire Protection Principle Backflow Preventers (RFP)	312.10.2
5015—98	Performance Requirements for Testing Double Check Valve Backflow Prevention Assembly (DCVA)	312.10.2
5020—98	Performance Requirements for Testing Pressure Vacuum Breaker Assembly (PVBA)	312.10.2
5047—98	Performance Requirements for Testing Reduced Pressure Detector Fire Protection Backflow Prevention Assemblies (RPDA)	312.10.2
5048—98	Performance Requirements for Testing Double Check Valve Detector Assembly (DCDA)	312.10.2
5052—98	Performance Requirements for Testing Hose Connection Backflow Preventers	312.102
5056—98	Performance Requirements for Testing Spill Resistant Vacuum Breaker	312.10.2

ASTM

ASTM International
100 Barr Harbor Drive
West Conshohocken, PA 19428-2959

Standard Reference Number	Title	Referenced in code section number
A 53/A 53M—06a	Specification for Pipe, Steel, Black and Hot-dipped, Zinc-coated Welded and Seamless	Table 605.3, Table 605.4, Table 702.1
A 74—06	Specification for Cast-iron Soil Pipe and Fittings	Table 702.1, Table 702.2, Table 702.3, Table 702.4, 708.2, 708.7, Table 1102.4, Table 1102.5, Table 1102.7
A 312/A 312M—06	Specification for Seamless and Welded Austenitic Stainless Steel Pipes	Table 605.3, Table 605.4, Table 605.5, 605.23.2
A 733—03	Specification for Welded and Seamless Carbon Steel and Austenitic Stainless Steel Pipe Nipples	Table 605.8
A 778—01	Specification for Welded Unannealed Austenitic Stainless Steel Tubular Products	Table 605.3, Table 605.4, Table 605.5
A 888—07a	Specification for Hubless Cast-iron Soil Pipe and Fittings for Sanitary and Storm Drain, Waste, and Vent Piping Application	Table 702.1, Table 702.2, Table 702.3, Table 702.4, 708.7, Table 1102.4, Table 1102.5, Table 1102.7
B 32—04	Specification for Solder Metal	605.14.3, 605.15.4, 705.9.3, 705.10.3
B 42—02e01	Specification for Seamless Copper Pipe, Standard Sizes	Table 605.3, Table 605.4, Table 702.1
B 43—98(2004)	Specification for Seamless Red Brass Pipe, Standard Sizes	Table 605.3, Table 605.4, Table 702.1
B 75—02	Specification for Seamless Copper Tube	Table 605.3, Table 605.4, Table 702.1, Table 702.2, Table 702.3, Table 1102.4
B 88—03	Specification for Seamless Copper Water Tube	Table 605.3, Table 605.4, Table 702.1, Table 702.2, Table 702.3, Table 1102.4
B 152/B 152M—06a	Specification for Copper Sheet, Strip Plate and Rolled Bar	402.3, , 417.5.2.4, 425.3.3, 902.2
B 251—02e01	Specification for General Requirements for Wrought Seamless Copper and Copper-alloy Tube	Table 605.3, Table 605.4, Table 702.1, Table 702.2, Table 702.3, Table 1102.4
B 302—02	Specification for Threadless Copper Pipe, Standard Sizes	Table 605.3, Table 605.4, Table 702.1
B 306—02	Specification for Copper Drainage Tube (DWV)	Table 702.1, Table 702.2, Table 1102.4
B 447—07	Specification for Welded Copper Tube	Table 605.3, Table 605.4
B 687—99(2005)e01	Specification for Brass, Copper and Chromium-plated Pipe Nipples	Table 605.8
B 813—00e01	Specification for Liquid and Paste Fluxes for Soldering of Copper and Copper Alloy Tube	605.14.3, 605.15.4, 705.9.3, 705.10.3
B 828—02	Practice for Making Capillary Joints by Soldering of Copper and Copper Alloy Tube and Fittings	605.14.3, 605.15.4, 705.9.3, 705.10.3
C 4—04e01	Specification for Clay Drain Tile and Perforated Clay Drain Tile	Table 702.3, Table 1102.4, Table 1102.5
C 14—07	Specification for Nonreinforced Concrete Sewer, Storm Drain and Culvert Pipe	Table 702.3, Table 1102.4
C 76—07	Specification for Reinforced Concrete Culvert, Storm Drain and Sewer Pipe	Table 702.3, Table 1102.4
C 296—(2004)e01	Specification for Asbestos-cement Pressure Pipe	Table 605.3
C 425—04	Specification for Compression Joints for Vitrified Clay Pipe and Fittings	705.15, 705.19
C 428—97(2006)	Specification for Asbestos-cement Nonpressure Sewer Pipe	Table 702.2, Table 702.3, Table 702.4, Table 1102.4
C 443—05a	Specification for Joints for Concrete Pipe and Manholes, Using Rubber Gaskets	705.6, 705.19
C 508—(2004)	Specification for Asbestos-cement Underdrain Pipe	Table 1102.5
C 564—04a	Specification for Rubber Gaskets for Cast-iron Soil Pipe and Fittings	705.5.2, 705.5.3, 705.19, Table 1102.4
C 700—07	Specification for Vitrified Clay Pipe, Extra Strength, Standard Strength, and Perforated	Table 702.3, 702.4, Table 1102.4, Table 1102.5
C 1053—00(2005)	Specification for Borosilicate Glass Pipe and Fittings for Drain, Waste, and Vent (DWV) Applications	Table 702.1, Table 702.4
C 1173—06	Specification for Flexible Transition Couplings for Underground Piping System	705.2.1, 705.7.1, 705.14.1, 705.15, 705.16.1, 705.19
C 1277—06	Specification for Shielded Coupling Joining Hubless Cast-iron Soil Pipe and Fittings	705.5.3
C 1440—03	Specification for Thermoplastic Elastomeric (TPE) Gasket Materials for Drain, Waste,and Vent (DWV), Sewer, Sanitary and Storm Plumbing Systems	705.19
C 1460—04	Specification for Shielded Transition Couplings for Use with Dissimilar DWV Pipe and Fittings Above Ground	705.19
C 1461—06	Specification for Mechanical Couplings Using Thermoplastic Elastomeric (TPE) Gaskets for Joining Drain, Waste and Vent (DWV) Sewer, Sanitary and Storm Plumbing Systems for Above and Below Ground Use	705.19
C 1540—04	Specification for Heavy Duty Shielded Couplings Joining Hubless Cast-iron Soil Pipe and Fittings	705.5.3
C 1563—04	Standard Test Method for Gaskets for Use in Connection with Hub and Spigot Cast Iron Soil Pipe and Fittings for Sanitary Drain, Waste, Vent and Storm Piping Applications.	705.5.2

ASTM—continued

Standard	Title	Referenced in
D 1527—99(2005)	Specification for Acrylonitrile-Butadiene-Styrene (ABS) Plastic Pipe, Schedules 40 and 80	Table 605.3
D 1785—06	Specification for Poly (Vinyl Chloride) (PVC) Plastic Pipe, Schedules 40, 80 and 120	Table 605.3
D 1869—95(2005)	Specification for Rubber Rings for Asbestos-cement Pipe	605.11, 605.24, 705.3, 705.19
D 2235—04	Specification for Solvent Cement for Acrylonitrile-Butadiene-Styrene (ABS) Plastic Pipe and Fittings	605.10.2, 705.2.2, 705.7.2
D 2239—03	Specification for Polyethylene (PE) Plastic Pipe (SIDR-PR) Based on Controlled Inside Diameter	Table 605.3
D 2241—05	Specification for Poly (Vinyl Chloride) (PVC) Pressure-rated Pipe (SDR-Series)	Table 605.3
D 2282—(2005)99e01	Specification for Acrylonitrile-Butadiene-Styrene (ABS) Plastic Pipe (SDR-PR)	Table 605.3
D 2464—06	Specification for Threaded Poly (Vinyl Chloride) (PVC) Plastic Pipe Fittings, Schedule 80	Table 605.5, Table 1102.7
D 2466—06	Specification for Poly (Vinyl Chloride) (PVC) Plastic Pipe Fittings, Schedule 40	Table 605.5, Table 1102.7
D 2467—06	Specification for Poly (Vinyl Chloride) (PVC) Plastic Pipe Fittings, Schedule 80	Table 605.5, Table 1102.7
D 2468—96a	Specification for Acrylonitrile-Butadiene-Styrene (ABS) Plastic Pipe Fittings, Schedule 40	Table 605.5, Table 1102.7
D 2564—04e01	Specification for Solvent Cements for Poly (Vinyl Chloride) (PVC) Plastic Piping Systems	605.21.2, 705.8.2, 705.14.2
D 2609—02	Specification for Plastic Insert Fittings for Polyethylene (PE) Plastic Pipe	Table 605.5, Table 1102.7
D 2657—07	Practice for Heat Fusion-joining of Polyolefin Pipe and Fitting	605.19.2, 705.16.1
D 2661—06	Specification for Acrylonitrile-Butadiene-Styrene (ABS) Schedule 40 Plastic Drain, Waste, and Vent Pipe and Fittings	Table 702.1, Table 702.2, Table 702.3, Table 702.4, 705.2.2, 705.7.2, Table 1102.4, Table 1102.7
D 2665—07	Specification for Poly (Vinyl Chloride) (PVC) Plastic Drain, Waste, and Vent Pipe and Fittings	Table 702.1, Table 702.2, Table 702.3, Table 702.4, Table 1102.4, Table 1102.7
D 2672—96a(2003)	Specification for Joints for IPS PVC Pipe Using Solvent Cement	Table 605.3
D 2683—04	Standard Specification for Socket-type Polyethylene fittings for Outside Diameter-controlled Polyethylene Pipe and Tubing	Table 605.5
D 2729—04e01	Specification for Poly (Vinyl Chloride) (PVC) Sewer Pipe and Fittings	Table 1102.5
D 2737—03	Specification for Polyethylene (PE) Plastic Tubing	Table 605.3
D 2751—05	Specification for Acrylonitrile-Butadiene-Styrene (ABS) Sewer Pipe and Fittings	Table 702.3, Table 702.4, Table 1102.7
D 2846/D 2846M—06	Specification for Chlorinated Poly (Vinyl Chloride) (CPVC) Plastic Hot and Cold Water Distribution Systems	Table 605.3, Table 605.4, Table 605.5, 605.16.2
D 2855—96(2002)	Standard Practice for Making Solvent-cemented Joints with Poly (Vinyl Chloride) (PVC) Pipe and Fittings	605.22.2, 705.8.2, 705.14.2
D 2949—01ae01	Specification for 3.25-in Outside Diameter Poly (Vinyl Chloride) (PVC) Plastic Drain, Waste, and Vent Pipe and Fittings	Table 702.1, Table 702.2, Table 702.3, Table 702.4
D 3034—06	Specification for Type PSM Poly (Vinyl Chloride) (PVC) Sewer Pipe and Fittings	Table 702.3, Table 702.4, Table 1102.7, Table 1102.4
D 3035-03	Standard Specification for Polyethylene (PE) Plastic Pipe (DR-PR) Based on Controlled Outside Diameter	Table 605.3
D 3139—98(2005)	Specification for Joints for Plastic Pressure Pipes Using Flexible Elastomeric Seals	605.10.1, 605.22.1
D 3212—96a(2003)e01	Specification for Joints for Drain and Sewer Plastic Pipes Using Flexible Elastomeric Seals	705.2.1, 705.8.1, 705.14.1, 705.16.2
D 3261—03	Standard Specification for Butt Heat Fusion Polyethylene (PE) Plastic fittings for Polyethylene (PE) Plastic Pipe and Tubing	Table 605.5
D 3311—06a	Specification for Drain, Waste and Vent (DWV) Plastic Fittings Patterns	Table 1102.7
D 4068—01	Specification for Chlorinated Polyethlene (CPE) Sheeting for Concealed Water-containment Membrane	417.5.2.2
D 4551—96(2001)	Specification for Poly (Vinyl Chloride) (PVC) Plastic Flexible Concealed Water-containment Membrane	417.5.2.1
F 405—05	Specification for Corrugated Polyethylene (PE) Tubing and Fittings	Table 1102.5
F 409—02	Specification for Thermoplastic Accessible and Replaceable Plastic Tube and Tubular Fittings	424.1.2, Table 1102.7
F 437—06	Specification for Threaded Chlorinated Poly (Vinyl Chloride) (CPVC) Plastic Pipe Fittings, Schedule 80	Table 605.5
F 438—04	Specification for Socket-type Chlorinated Poly (Vinyl Chloride) (CPVC) Plastic Pipe Fittings, Schedule 40	Table 605.5
F 439—06	Standard Specification for Chlorinated Poly (Vinyl Chloride) (CPVC) Plastic Pipe Fittings, Schedule 80	Table 605.5
F 441/F 441M—02	Specification for Chlorinated Poly (Vinyl Chloride) (CPVC) Plastic Pipe, Schedules 40 and 80	Table 605.3, Table 605.4
F 442/F 442M—99(2005)	Specification for Chlorinated Poly (Vinyl Chloride) (CPVC) Plastic Pipe (SDR-PR)	Table 605.3, Table 605.4
F 477—07	Specification for Elastomeric Seals (Gaskets) for Joining Plastic Pipe	605.24, 705.19
F 493—04	Specification for Solvent Cements for Chlorinated Poly (Vinyl Chloride) (CPVC) Plastic Pipe and Fittings	605.16.2
F 628—06e01	Specification for Acrylonitrile-Butadiene-Styrene (ABS) Schedule 40 Plastic Drain, Waste, and Vent Pipe with a Cellular Core	Table 702.1, Table 702.2, Table 702.3, Table 702.4, 705.2.2, 705.7.2, Table 1102.4, Table 1102.7

ASTM—continued

F 656—02	Specification for Primers for Use in Solvent Cement Joints of Poly (Vinyl Chloride) (PVC) Plastic Pipe and Fittings	605.22.2, 705.8.2, 705.14.2
F 714—06a	Specification for Polyethylene (PE) Plastic Pipe (SDR-PR) Based on Outside Diameter	Table 702.3
F 876—06	Specification for Cross-linked Polyethylene (PEX) Tubing	Table 605.3, Table 605.4
F 877—07	Specification for Cross-linked Polyethylene (PEX) Plastic Hot and Cold Water Distribution Systems	Table 605.3, Table 605.4, Table 605.5
F 891—04	Specification for Coextruded Poly (Vinyl Chloride) (PVC) Plastic Pipe with a Cellular Core	Table 702.1 Table 702.2, Table 702.3, Table 1102.4, Table 1102.5, Table 1102.7
F 1055—98(2006)	Standard Specification for Electrofusion Type Polyethylene Fittings for Outside Diameter Controlled Polyethylene Pipe and Tubing	Table 605.5
F 1281—07	Specification for Cross-linked Polyethylene/Aluminum/ Cross-linked Polyethylene (PEX-AL-PEX) Pressure Pipe	Table 605.3, Table 605.4, Table 605.5, 605.21.1
F 1282—06	Specification for Polyethylene/Aluminum/Polyethylene (PE-AL-PE) Composite Pressure Pipe	Table 605.3, Table 605.4, Table 605.5, 605.21.1
F 1412—01e01	Specification for Polyolefin Pipe and Fittings for Corrosive Waste Drainage	Table 702.1, Table 702.2, Table 702.4, 705.17.1
F 1488—03	Specification for Coextruded Composite Pipe	Table 702.1, Table 702.2, Table 702.3
F 1673—04	Polyvinylidene Fluoride (PVDF) Corrosive Waste Drainage Systems	Table 702.1, Table 702.2, Table 702.3, Table 702.4, 705.18.1
F 1807—07	Specification for Metal Insert Fittings Utilizing a Copper Crimp Ring for SDR9 Cross-linked Polyethylene (PEX) Tubing	Table 605.5
F 1866—07	Specification for Poly (Vinyl Chloride) (PVC) Plastic Schedule 40 Drainage and DWV Fabricated Fittings	Table 702.4, Table 1102.7
F 1960—07	Specification for Cold Expansion Fittings with PEX Reinforcing Rings for use with Cross-linked Polyethylene (PEX) Tubing	Table 605.5
F 1974—04	Specification for Metal Insert Fittings for Polyethylene/Aluminum/Polyethylene and Cross-linked Polyethylene/Aluminum/Cross-linked Polyethylene Composite Pressure Pipe	Table 605.5, 605.21.1
F 1986—01(2006)	Specification for Multilayer Pipe, Type 2, Compression Fittings and Compression Joints for Hot and Cold Drinking Water Systems	Table 605.3, Table 605.4, Table 605.5
F 2080—05	Specifications for Cold-expansion Fittings with Metal Compression-sleeves for Cross-linked Polyethylene (PEX) Pipe	Table 605.5
F 2098—04e01	Standard specification for Stainless Steel Clamps for Securing SDR9 Cross-linked Polyethylene (PEX) Tubing to Metal Insert Fittings	Table 605.5
F 2159—05	Specification for Plastic Insert Fittings Utilizing a Copper Crimp Ring for SDR9 Cross-linked Polyethylene (PEX) Tubing	Table 605.5
F 2262—03	Specification for Cross-linked Polyethylene/Aluminum/Cross-linked Polyethylene Tubing OD Controlled SDR9	Table 605.3, Table 605.4
F 2306/F 2306M-05	12" to 60" Annular Corrugated Profile-wall Polyethylene (PE) Pipe and Fittings for Gravity Flow Storm Sewer and Subsurface Drainage Applications	Table 1102.4, Table 1102.7
F 2389—06	Specification for Pressure-rated Polypropylene (PP) Piping Systems	Table 605.3, Table 605.4, Table 605.5, 605.20.1
F 2434—05	Standard Specification for Metal Insert Fittings Utilizing a Copper Crimp Ring for SDR9 Cross-linked Polyethylene (PEX) Tubing and SDR9 Cross-linked Polyethylene/ Aluminum/Cross-linked Polyethylene (PEX AL-PEX) Tubing	Table 605.5

AWS

American Welding Society
550 N.W. LeJeune Road
Miami, FL 33126

Standard Reference Number	Title	Referenced in code section number
A5.8—04	Specifications for Filler Metals for Brazing and Braze Welding	605.12.1, 605.14.1, 605.15.1, 705.4.1, 705.9.1, 705.10.1

AWWA

American Water Works Association
6666 West Quincy Avenue
Denver, CO 80235

Standard Reference Number	Title	Referenced in code section number
C104—98	Standard for Cement-mortar Lining for Ductile-iron Pipe and Fittings for Water	605.3, 605.5
C110—/A21.10—03	Standard for Ductile-iron and Gray-iron Fittings, 3 Inches through 48 Inches, for Water	Table 605.5, Table 702.4, Table 1102.7
C111—00	Standard for Rubber-gasket Joints for Ductile-iron Pressure Pipe and Fittings	605.13
C115/A21.15—99	Standard for Flanged Ductile-iron Pipe with Ductile-iron or Gray-iron Threaded Flanges	Table 605.3, Table 605.4
C151/A21.51—02	Standard for Ductile-iron Pipe, Centrifugally Cast for Water	Table 605.3, Table 605.4
C153—00/A21.53—00	Standard for Ductile-iron Compact Fittings for Water Service	Table 605.5
C510—00	Double Check Valve Backflow Prevention Assembly	Table 608.1, 608.13.7
C511—00	Reduced-pressure Principle Backflow Prevention Assembly	Table 608.1, 608.13.2, 608.16.2
C651—99	Disinfecting Water Mains	610.1
C652—02	Disinfection of Water-storage Facilities	610.1

CISPI

Cast Iron Soil Pipe Institute
5959 Shallowford Road, Suite 419
Chattanooga, TN 37421

Standard Reference Number	Title	Referenced in code section number
301—04a	Specification for Hubless Cast-iron Soil Pipe and Fittings for Sanitary and Storm Drain, Waste and Vent Piping Applications	Table 702.1, Table 702.2, Table 702.3, Table 702.4, 708.7, Table 1102.4, Table 1102.5, Table 1102.7
310—04	Specification for Coupling for Use in Connection with Hubless Cast-iron Soil Pipe and Fittings for Sanitary and Storm Drain, Waste and Vent Piping Applications	705.5.3

CSA

Canadian Standards Association
5060 Spectrum Way.
Mississauga, Ontario, Canada L4W 5N6

Standard Reference Number	Title	Referenced in code section number
B45.1—02	Ceramic Plumbing Fixtures	408.1, 416.1, 418.1, 419.1, 420.1
B45.2—02	Enameled Cast-iron Plumbing Fixtures	407.1, 415.1, 416.1, 418.1
B45.3—02	Porcelain Enameled Steel Plumbing Fixtures	407.1, 416.1, 418.1
B45.4—02	Stainless-steel Plumbing Fixtures	415.1, 416.1, 418.1, 420.1
B45.5—02	Plastic Plumbing Fixtures	407.1, 416.2, 417.1, 419.1, 420.1, 421.1
B45.9—99	Macerating Systems and Related Components	712.4.1
B64.1.2—01	Vacuum Breakers, Pressure Type (PVB)	Table 608.1, 608.13.5
B64.2.1—01	Vacuum Breakers, Hose Connection Type (HCVB) with Manual Draining Feature	Table 608.1, 608.13.6
B64.2.1.1—01	Vacuum Breakers, Hose Connection Dual Check Type (HCDVB)	Table 608.1, 608.13.6
B64.4.1—01	Backflow Preventers, Reduced Pressure Principle Type for Fire Sprinklers (RPF)	Table 608.1, 608.13.2
B64.5—01	Backflow Preventers, Double Check Type (DCVA)	Table 608.1, 608.13.7
B64.5.1—01	Backflow Preventers, Double Check Type for Fire Systems (DCVAF)	Table 608.1 608.13.7
B64.6—01	Backflow Preventers, Dual Check Valve Type (DuC)	605.3.1, Table 608.1
B64.7—94	Vacuum Breakers, Laboratory Faucet Type (LFVB	Table 608.1, 608.13.6
B64.10/B64.10.1—01	Manual for the Selection and Installation of Backflow Prevention Devices/Manual for the Maintenance and Field Testing of Backflow Prevention Devices	312.10.2
B79—94(2000)	Floor, Area and Shower Drains, and Cleanouts for Residential Construction	412.1
B125—01	Plumbing Fittings	424.4, 424.6, 425.4
B125.3—2005	Plumbing Fittings	416.5, 424.5, 425.3.1, Table 608.1
B137.1—02	Polyethylene Pipe, Tubing and Fittings for Cold Water Pressure Services	Table 605.3
B137.2—02	PVC Injection-moulded Gasketed Fittings for Pressure Applications	Table 605.5, Table 1102.7
B137.3—02	Rigid Poly (Vinyl Chloride) (PVC) Pipe for Pressure Applications	Table 605.3, Table 605.4, Table 605.5, 605.22.2, 705.8.2, 705.14.2
B137.5—02	Cross-linked Polyethylene (PEX) Tubing Systems for Pressure Applications—with Revisions through September 1992	Table 605.3, Table 605.4, Table 605.5
B137.6—02	CPVC Pipe, Tubing and Fittings for Hot and Cold Water Distribution Systems—with Revisions through May 1986	Table 605.3, Table 605.4
B137.11—02	Polypropylene (PP-R) Pipe and Fittings for Pressure Applications	Table 605.3, Table 605.4, Table 605.5
B181.1—02	ABS Drain, Waste and Vent Pipe and Pipe Fittings	Table 702.1, Table 702.2, Table 702.3, Table 702.4, 705.2.2, 705.7.2, 715.2, Table 1102.4, Table 1102.7
B181.2—02	PVC Drain, Waste, and Vent Pipe and Pipe Fittings—with Revisions through December 1993	Table 702.1 Table 702.2, 705.8.2, 705.14.2, 715.2
B182.1—02	Plastic Drain and Sewer Pipe and Pipe Fittings	705.8.2, 705.14.2, Table 1102.4
B182.2—02	PVC Sewer Pipe and Fittings (PSM Type)	Table 702.3, Table 1102.4, Table 1102.5
B182.4—02	Profile PVC Sewer Pipe and Fittings	Table 702.3, Table 1102.4, Table 1102.5
B182.6—02	Profile Polyethylene Sewer Pipe and Fittings for Leak-proof Sewer Applications	Table 1102.5
B182.8—02	Profile Polyethylene Storm Sewer and Drainage Pipe and Fittings	Table 1102.5
CAN/CSA-A257.1M—92	Circular Concrete Culvert, Storm Drain, Sewer Pipe and Fittings	Table 702.3, Table 1102.4
CAN/CSA-A257.2M—92	Reinforced Circular Concrete Culvert, Storm Drain, Sewer Pipe and Fittings	Table 702.3, Table 1102.4
CAN/CSA-A257.3M—92	Joints for Circular Concrete Sewer and Culvert Pipe, Manhole Sections and Fittings Using Rubber Gaskets	705.6, 705.19
CAN/CSA-B64.1.1—01	Vacuum Breakers, Atmospheric Type (AVB)	425.2, Table 608.1, 608.13.6
CAN/CSA-B64.2—01	Vacuum Breakers, Hose Connection Type (HCVB)	Table 608.1, 608.13.6
CAN/CSA-B64.2.2—01	Vacuum Breakers, Hose Connection Type (HCVB) with Automatic Draining Feature	Table 608.1, 608.13.6
CAN/CSA-B64.3—01	Backflow Preventers, Dual Check Valve Type with Atmospheric Port (DCAP)	Table 608.1, 608.13.3, 608.16.2
CAN/CSA-B64.4—01	Backflow Preventers, Reduced Pressure Principle Type (RP)	Table 608.1, 608.13.2, 608.16.2
CAN/CSA-B64.10—01	Manual for the Selection, Installation, Maintenance and Field Testing of Backflow Prevention Devices	312.10.2
CAN/CSA-B137.9—02	Polyethylene/Aluminum/Polyethylene Composite Pressure Pipe Systems	Table 605.3, Table 605.5, 605.21.1
CAN/CSA-B137.10M—02	Cross-linked Polyethylene/Aluminum/Polyethylene Composite Pressure Pipe Systems	Table 605.3, Table 605.4, Table 605.5, 605.21.1
CAN/CSA-B181.3—02	Polyolefin Laboratory Drainage Systems	Table 702.1, Table 702.2, Table 702.4, 705.17.1
CAN/CSA-B182.4—02	Profile PVC Sewer Pipe and Fittings	Table 702.3, Table 1102.4, Table 1102.5
CAN/CSA-B602—02	Mechanical Couplings for Drain, Waste and Vent Pipe and Sewer Pipe	705.2.1, 705.5.3, 705.6, 705.7.1, 705.14.1, 705.15, 705.16.2, 705.19

ICC

International Code Council, Inc.
500 New Jersey Ave, NW
6th Floor
Washington, DC 20001

Standard Reference Number	Title	Referenced in code section number
IBC—09	International Building Code®	201.3, 305.4, 307.1, 307.2, 307.3, 308.2, 309.1,310.1, 310.3, 403.1, Table 403.1, 404.1, 407.3, 417.6, 502.6, 606.5.2, 1106.5
IEBC—09	International Existing Building Code	101.2
IECC—09	International Energy Conservation Code	313.1, 607.2, 607.2.1
IFC—09	International Fire Code®	201.3, 1201.1
IFGC—09	International Fuel Gas Code®	101.2, 201.3, 502.1
IMC—09	International Mechanical Code®	201.3, 307.6, 310.1, 422.9, 502.1, 612.1, 1202.1
IPSDC—09	International Private Sewage Disposal Code®	701.2
IRC—09	International Residential Code	101.2

ISEA

International Safety Equipment Association
1901 N. Moore Street, Suite 808
Arlington, VA 22209

Standard Reference Number	Title	Referenced in code section number
Z358.1—98	Emergency Eyewash and Shower Equipment	411.1

NFPA

National Fire Protection Association
1 Batterymarch Park
Quincy, MA 02169-7471

Standard Reference Number	Title	Referenced in code section number
50—01	Bulk Oxygen Systems at Consumer Sites	1203.1
51—07	Design and Installation of Oxygen-fuel Gas Systems for Welding, Cutting and Allied Processes	1203.1
70—08	National Electric Code	502.1, 504.3, 1113.1.3
99C—05	Gas and Vacuum Systems	1202.1

NSF

NSF International
789 Dixboro Road
Ann Arbor, MI 48105

Standard Reference Number	Title	Referenced in code section number
3—2007	Commercial Warewashing Equipment	409.1
14—2007	Plastic Piping System Components and Related Materials	303.3, 611.3
18—2007	Manual Food and Beverage Dispensing Equipment	426.1
42—2007e	Drinking Water Treatment Units—Aesthetic Effects	611.1, 611.3
44—2004	Residential Cation Exchange Water Softeners	611.1, 611.3
53—2007	Drinking Water Treatment Units—Health Effects	611.1, 611.3
58—2006	Reverse Osmosis Drinking Water Treatment Systems	611.2
61—2007a	Drinking Water System Components—Health Effects	410.1, 424.1, 605.3, 605.4, 605.5, 605.7, 611.3, 611.3
62—2004	Drinking Water Distillation Systems	611.1

PDI

Plumbing and Drainage Institute
800 Turnpike Street, Suite 300
North Andover, MA 01845

Standard Reference Number	Title	Referenced in code section number
G101(2003)	Testing and Rating Procedure for Grease Interceptors with Appendix of Sizing and Installation Data	1003.3.4

UL

Underwriters Laboratories, Inc.
333 Pfingsten Road
Northbrook, IL 60062-2096

Standard Reference Number	Title	Referenced in code section number
UL508—99	Industrial Control Equipment with Revision through July 2005	314.2.3

Index